Macmillan Master Series

Accounting
Arabic
Astronomy
Australian History
Background to Business
Banking
Basic Management
Biology
British Politics
Business Communication
Business Law
Business Microcomputing
C Programming
Catering Science
Catering Theory
Chemistry
COBOL Programming
Commerce
Computer Programming
Computers
Economic and Social History
Economics
Electrical Engineering
Electronics
English as a Foreign Language
English Grammar
English Language
English Literature
Financial Accounting
French 1
French 2
German 1

German 2
Hairdressing
Human Biology
Italian 1
Italian 2
Japanese
Keyboarding
Marketing
Mathematics
Modern British History
Modern European History
Modern World History
Nutrition
Office Practice
Pascal Programming
Philosophy
Physics
Practical Writing
Principles of Accounts
Psychology
Restaurant Service
Science
Social Welfare
Sociology
Spanish 1
Spanish 2
Spreadsheets
Statistics
Statistics with your Microcomputer
Study Skills
Typewriting Skills
Word Processing

Mastering

Science

Robert Barrass

M
MACMILLAN

First published 1991

10 9 8 7 6 5 4 3 2
00 99 98 97 96 95 94 93 92 91

Published by
MACMILLAN EDUCATION LTD
Houndmills, Basingstoke, Hampshire RG21 2XS
and London
Companies and representatives
throughout the world

Printed in Hong Kong

British Library Cataloguing in Publication Data
Barrass, Robert
Mastering science
1. Science
500
ISBN 0 – 333 – 49985 – 9 pbk
      0 – 333 – 49986 – 7 export edition

To *Wilfred Biggs*
1929–1987
a teacher of science,
and of science teachers

Contents

⬡ Preface

This book can be read as a whole or used for reference by anyone who requires a concise introduction to science and to the way scientists work. It is especially suitable for those who are preparing for an examination in science.

No previous knowledge of science is assumed. If you are working alone and have not studied science before, start at the beginning and work through to the end. Study the illustrations as well as the text. You will find the information needed for a proper understanding of each topic not only on the page where the topic is considered but also on the preceding pages, so cross references are provided as reminders.

Science is presented in historical perspective, as an increasing body of knowledge and as an aspect of human endeavour. Due reference is made to the relevance of science to our everyday lives — to the economic importance, environmental impact and social consequences of many discoveries.

Simple practical investigations are included to encourage careful observation, precise measurement, accurate recording, interpretation of evidence, and preparation of concise written reports. Most of these investigations can be undertaken with readily available materials.

Assessment, by a teacher, of practical work and communication skills is an essential part of most courses in science. Readers preparing for examinations, especially if they have not previously studied science at school or college, are advised to enrol for a science course that includes laboratory investigations.

Warning. Investigations for which special apparatus or chemicals are needed should be undertaken only in a laboratory and only in the presence of a suitably qualified and experienced teacher.

In this book only essential technical terms are used and when a new term is introduced (see Index) it is clearly defined or its meaning is made clear in a labelled drawing or diagram. The use of mathematics is kept to a minimum but some calculations are set out clearly, as worked examples, where they are needed.

The Questions (Appendix 1) are similar to those set in science examinations and the Further reading (Appendix 3) is for those who would like to prepare for a more advanced course in a science subject or would simply like to know more about science.

Robert Barrass

⬡ Acknowledgements

I started to plan an introduction to science in 1979. Between 1983 and 1985 I was helped in preparing specimen chapters for a publisher by my friend and colleague Wilfred Biggs to whom this book is dedicated. I benefited greatly from discussions with him and, when I started work on this book in 1989, from the use of papers passed on to me by his wife Rosemary.

I thank Jonathan Barrass for reading several chapters in typescript, particularly those dealing with electricity and electronics; Dennis Wheeler for reading the chapter on climate and weather; and Stephen Foster who read the whole typescript for my publishers. The book has been improved as a result of their suggestions. I also thank my wife Ann for her help.

I shall be pleased to receive suggestions for improving future editions.

Robert Barrass
Sunderland Polytechnic
March 1990

1 Science for you

1.1 Science and scientists

Science is the study of everything in the universe — including the planet Earth, its atmosphere and every living thing.

The word science means knowledge. People who use their knowledge and understanding of science in their work are called scientists.

Table 1.1 *Some science subjects*

Sciences	What scientists study
Agriculture	Food production on land
Anthropology	People and their origins
Astronomy	Heavenly bodies, including the moon, planets and stars
Biochemistry	The chemistry of living things
Biology	Living things
Chemistry	Substances and how they combine
Geology	Earth's crust
Physics	Materials and energy

This book is an introduction to:
1 The way scientists work
2 The knowledge and understanding gained as a result of their work
3 How people have used this knowledge
4 Some economic and social consequences of these uses

For example, as you study science you will learn how scientists have helped to prevent or cure many diseases of crop plants, farm animals and people; to increase food production from crop plants and farm animals; and to improve the conditions in which many people live and work. These are some of the economic and social uses of science.

There are many careers for which the study of science is a good beginning. Later you may take a more advanced course in one or more science subjects (see Table 1.1). But discoveries in science have affected our lives so much that even if you do not become a professional scientist you will still find it useful to have some knowledge of science. Also, knowing how scientists investigate problems will help you to overcome difficulties you encounter in studying other subjects — and in the rest of your life.

1.2 Our cultural evolution

A culture is a way people live together — at any time and in any place — based on their knowledge, skills, beliefs and other influences. The word evolution simply means change. Different periods in our cultural evolution are named according to the materials people had learnt to use (see Table 1.2).

Until about 300 000 years ago people probably had little more effect on their surroundings than did any other animal. Since then, by using fire, people have affected the growth of vegetation over large areas of the world.

The start of cultivation, about 10 000 years ago, has also had lasting effects on the world's vegetation and soils. As time has passed, more and more land has been used for food production, yielding more food than had previously been available from fishing, hunting wild animals and collecting wild plants. People were able to settle in one place, grow and store food, and feed more people throughout the year.

In some places small settlements grew into villages, and some villages into towns and cities in which most people were not directly involved in food production. These people specialised in different kinds of work — a division of labour.

The words civilisation and citizen are based on two Latin words *civitas* (the state) and *civis* (the citizen). In the social life of any community each citizen had rights and responsibilities. These were accepted by most members of the community and with the development of writing they could be recorded.

Communication

The development of writing, between 5000 and 3000 years ago, made it possible for people to record observations, ideas, opinions and beliefs. Only people have developed both spoken and written languages.

People can let us know what they are thinking by facial expressions, gestures and speech. We can benefit from the experience of others, we can argue and discuss. Also, as a result of drawing, writing, and other methods of recording, we know something of how people lived in the past — and we can convey information and ideas to future generations.

Increasing knowledge has allowed changes in the way people live (see Table 1.2).

Astronomy and mathematics, closely associated with religion, were important in early civilisations that developed between 5000–3000 years ago in, for example, China, India, Mesopotamia, Egypt, and Central America. We still use the Hindu-Arabic numbers (0 to 9) and the

Table 1.2   *Our cultural evolution*

Materials used	How people lived
Old stone age	*Starting about 1 500 000 years ago* For more than a million years people hunted or fished, and gathered food*. They used crude tools of wood, bone and stone. 300 000 years ago people were using fire.
New stone age	*Starting about 10 000 years ago* The start of the agricultural revolution. As well as hunting, fishing and collecting wild plants, people started to grow food. They had already domesticated some kinds of animals. People made carefully shaped tools and weapons of stone.
Bronze age	*Starting about 5000 years ago* Civilisations developed in many parts of the world. Bronze (a mixture of copper and tin) was used to make ornaments, tools and weapons. Animals were used as beasts of burden and wheeled vehicles were invented. Many people lived in towns and cities and were not directly involved in food production.
Iron age	*Starting about 2500 years ago* People started to make tools of iron. *About 500 years ago* The start of the scientific revolution. *About 250 years ago* The start of the industrial revolution, based on the invention of machines and their use in factories.
Age of steel	*About 125 years ago* In the last 50 years especially, as a result of discoveries in science, people are changing not only the conditions in which people live but also the conditions for all life on Earth.

Note. *A few cultures in isolated parts of the world are still living by hunting, fishing, and food gathering, much as people lived a million years ago. As a result of the influence of civilisations and the effects of world population growth, the days of such people are numbered.
This table is based on Barrass, R. *Human Biology Made Simple*.

numbering system based on 60 (for degrees, seconds and minutes), developed by civilisations in the Tigris and Euphrates valleys of Mesopotamia — where the wheel was in use 4500 years ago.

Beginning about 2500 years ago, Greek philosophers started to organise and record knowledge — including that derived from other civilisations. The word philosophy is from the Greek word *philosophia*, meaning love of knowledge. Aristotle wrote about all aspects of knowledge concerning nature — from astronomy to zoology. Much of his writing was based on careful observation and all his conclusions make clear his remarkable powers of reasoning. But his books did contain mistakes and misunderstandings, as well as accurate statements and correct deductions.

Unfortunately, many people considered everything Aristotle wrote to be true. As a result, the accepted way to study science was to read Aristotle. This discouraged scientific enquiry for the next 2000 years.

About 500 years ago, following the invention of the printing press, books by the Greek philosophers were printed and widely distributed. And as people published new observations and discoveries it became clear that some of the statements made by Greek philosophers were not correct.

1.3   The scientific method of investigation

There arose a new spirit of enquiry which could be said to mark the start of a scientific revolution. Scientists came to agree that speculation by itself was not enough, nor was the unsupported opinion of any expert (an authority on a subject). Instead they emphasised the need for accurate observation and precise measurement as the basis of science.

Observing and communicating

We are curious about things we can touch, taste, hear, smell and see. These things, which we recognise with our senses (see Table 10.1, page 103), are what scientists call observations.

We also use instruments so that we can make observations that would not otherwise be possible — things that are beyond the limits of our senses. For example, with a hand lens you can see details not visible to the unaided eye.

Here are some observations about flowering plants that you can repeat without special equipment.

Observations
1 Seeds are produced in fruits.
2 If seeds taken from a fruit are sown they will grow into plants if conditions are right.
3 These plants will be similar to those from which the fruit was taken and will produce similar fruit.

Interpretation
After making such observations about seeds from different kinds of plant, you might come to certain conclusions.
1 Seeds are produced only in fruits.

2 There is something in a seed that can develop into a plant.
3 All plants grown from seed are similar to those from which the fruit was taken and they produce similar seeds.

Prediction
On the basis of these conclusions, a scientist might predict that seeds from the fruit of a kind of plant he has not yet studied would produce more plants of that kind if they were sown in suitable soil.

Observation, interpretation and prediction are things people do well. These are part of the common sense way of working that scientists call the scientific method.

There are many investigations in this book. From them you will be able to learn for yourself many things about science and about the way scientists work. These investigations provide practice in:

1 Following instructions
2 Working safely
3 Observing carefully
4 Measuring precisely
5 Recording observations accurately
6 Analysing data (recorded observations) carefully
7 Interpreting the results of your analysis of data in the light of your own and other people's work
8 Preparing a written record of your work with appropriate reference to the work of others

Possible sources of error in any investigation include insufficient thought in planning the investigation; use of impure chemicals or faulty equipment; and insufficient care in reading instructions, observing, measuring, recording and analysing data, interpreting results, or in preparing a written account of the work.

Our ability to observe is also affected by the limitations of our senses and by the fact that we interpret observations in terms of our previous experience. This may result in their misinterpretation, as with optical illusions (see Figure 3.5). The influence of previous experience is also indicated by the fact that different scientists may interpret the same evidence and come to different conclusions.

Galileo Galilei (1564–1642), known as Galileo, was the first to emphasise the importance of careful observation, precise measurement and accurate recording as the basis of science. He also wrote clearly and forcefully and so gained widespread acceptance for the then new scientific method of investigation. In the cathedral at Pisa in Italy, Galileo is said to have watched lamps suspended from the ceiling as they moved to and fro, and to have counted the number of swings in a certain time. He repeated this observation and discovered something that you can find out for yourself.

Investigation 1.1   *Watching a pendulum swing* ─────

You will need about two metres of string with a small, heavy object tied at one end, a clock or watch with a sweeping second hand, and a notebook.

Proceed as follows:
1 Prepare a table (like Table 1.3) in your science notebook.
2 Tie the string to a fixed beam so the heavy object is free to swing. The heavy object (called the bob) and the string, together, make a *pendulum*.
3 Close doors and windows so there is no strong draught.
4 Allow the pendulum to hang (position B in Figure 1.1); then raise the bob to one side, keeping the string tight (position A in Figure 1.1). Release the bob to let the pendulum swing. The swing of a pendulum from its highest point on one side to its highest point on the other side (A through B to C in Figure 1.1) and then back again as far as it will go is called one *oscillation* of the pendulum.
5 Count the number of oscillations in one minute and record this number in your notebook (observation 1 in your table).
6 Repeat steps 4 and 5 several times and record in your table as observations 2, 3, and 4 and 5.

Table 1.3   *Recording numerical data in a table*

| | *Number of oscillations of pendulum in one minute* | | |
Observation	Investigation 1.1	Repeat	Experiment
1			
2			
3			
4			
5			
Total			
Average*			

* The *average* or *arithmetic mean* is calculated by dividing the sum of the items (the total) by the number of items (five in this table).

Keeping a record of your investigations. Scientists keep records as part of their work, to help them think, plan, observe, describe and remember. Keep a record of all your practical work in a special science notebook.

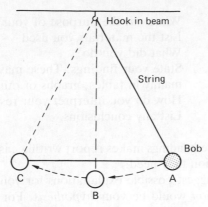

Figure 1.1 The swing of a pendulum

Note first the date and give each investigation a title. Then keep a concise note of exactly what you do. Write in carefully constructed sentences and prepare drawings (or diagrams like Figure 1.1) to ensure that your records cannot be misunderstood later.

Entering your observations in a table, as you make them, will help you to concentrate, to prepare complete and accurate records, to think about your work, and to arrange your thoughts. *Any observations recorded during an investigation are called data.* These are facts of any kind, recorded as words or numbers. Tables are particularly useful for recording numbers (numerical data) as you count or measure (as in Table 1.3). Counting and measuring enable scientists to be precise.

In Investigation 1.1 was each swing of the pendulum (a) longer, (b) as long as, or (c) shorter than the one before? Did the number of oscillations per minute (a) increase, (b) remain constant, or (c) decrease? Write two sentences in your notebook to summarise what you found in this investigation by observing and analysing your data to obtain your *results*.

The records you have made so far during your investigation are the kind a scientist would make in a laboratory notebook. Scientists also prepare accounts of their investigations for their own further reference and for publication so that other scientists can be informed of anything new. Communication is an essential part of the scientific method.

Make it a habit to prepare a written account of each of your own investigations. You will find that this adds to your understanding and is an aid to learning and remembering. Your teachers may use your written accounts of your investigations to help them assess your ability and understanding.

Start each account, or report, with a suitable title, your name and the date. Then use these headings to signpost the parts of your acount.

Introduction	What was the purpose of your investigation?
Materials	List the materials you used.
Methods	What did you do?
Results	State your findings. These may be presented mainly in tables, graphs or other diagrams.
Discussion	How do you interpret your results?
Conclusions	List any conclusions.

The use of these headings makes report writing easier and helps readers find the information they need.

If you could suggest possible explanations for your findings, each such *possible explanation* would be your *hypothesis*. For example, you might propose the hypothesis that as a pendulum swings the time taken for one oscillation is identical with the time taken for any other oscillation of the same pendulum. To test this hypothesis you could: (1) repeat your investigation with the same pendulum, recording your data below the heading *Repeat* in your table (Table 1.3), or (2) shorten the string and see what happens when this shorter pendulum swings to and fro. Record your data below the heading *Experiment* in your table (Table 1.3). In science *an experiment is an investigation in which a hypothesis is tested*.

Galileo confirmed his first observations — that a pendulum always takes as long for one oscillation as for another — experimentally. Because of this, it was soon realised that a pendulum could be used to regulate a clock. Pendulums are still used for this purpose. So Galileo's discovery in science has been used to solve a technical problem — accurate time keeping. Other methods of accurate time keeping have also been devised by scientists and are used in many clocks and watches instead of a pendulum.

The experimental method of testing hypotheses, which was devised by Galileo, is now accepted as part of the scientific method of investigation.

Scientists publish accounts of their investigations so that other scientists, and anyone else, can read about their work. Therefore, you can learn about science in two ways. One is by observation — most people remember best the things they have seen and touched but there is a limit to what you can find out for yourself, even in a lifetime. The other — and the quickest way to learn — is by listening to people talking about their work and by reading articles and books. By reading you can find out what scientists have selected from all they know about a subject. However, always remember that what you read is not necessarily true and even if true may not be a complete, balanced and unbiased account. Scientists should never have too much respect for the views of experts.

If you read books by different people you will find that there are many points upon which authors disagree. Some are questions of faith, concerning religious beliefs and cannot be tackled by observation and experiment. These are outside the scope of science. Others can be tackled by the scientific method of investigation. These are the questions to which scientists try to find answers.

② Earth in space, and time

2.1 Exploring Earth

Prehistoric peoples took an interest not only in their own world, which we call Earth, but also in the sun, moon and stars. They observed what seemed to be regular movements of the sun corresponding to the time of day and season of the year. They noted the regular phases of the moon, each lunar month, and that the highest tides were on the night of the new moon. They were impressed by these regular changes, which they could not understand but which seemed to regulate their lives, determining, for example, the time for planting and harvesting different crops. It is not surprising that the sun and moon were worshipped by some people.

In prehistoric times (before people kept written records), civilisations developed in different parts of the world. Travellers came to know the area around their own civilisation, and the land and sea routes between civilisations. Traders found their way across seas and over plains and deserts, where there were few if any landmarks, by noting the positions of certain groups of stars which we call constellations. For example, the plough (Figure 2.1) is visible from Earth's northern hemisphere throughout the year.

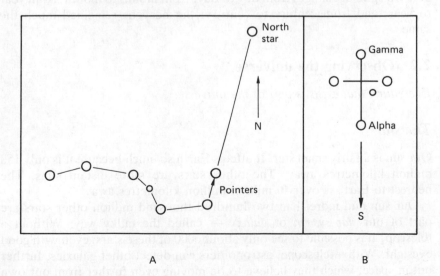

Figure 2.1 Constellations: (A) the plough and north star (or pole star); and (B) the southern cross.

Two stars in this constellation are called pointers because they are always in line with the north or pole star (Polaris), which indicates the direction in which north lies. Similarly, a group of stars that can be seen throughout the year from Earth's southern hemisphere is called the southern cross because it indicates the direction in which south lies (see Figure 2.1).

Early observers gave names to the things they observed, just as we do. Names already used in this chapter include sun, day, night, season, year, moon, month, time, tide, north and south. As emphasised in Chapter 1, communication is essential in science; and giving names to things is part of the development of any language. Before people can talk or write about anything they must give it a name. Observing things and naming them provides a basis for recording observations and so for the accumulation of knowledge.

Map-making is one way of putting observations together. Each map is a summary of observations. It shows, for example, the extent of land and sea, the positions and names of places, routes between places, and the direction in which north lies. A map drawn to scale helps those who wish either (a) to use these observations, or (b) to add new observations about Earth's surface.

The early observations of the sun, moon and stars were the beginning of the study of astronomy. These observations made possible the preparation of detailed records in the form of diagrams or maps of the appearance of the sky by day and by night and at different times of the year. From such studies it became clear that many things observed, including sunrise and sunset, and the phases of the moon, were regular and therefore predictable. In spite of changes from day to day, from month to month, from year to year, and from century to century, the heavens remained much the same.

2.2   Observing the universe

Everything that exists is part of the universe.

The stars

Our sun is a fairly small star. It affects Earth so much because it is only 150 million kilometres away. The other stars are more distant suns. The nearest to Earth is over 40 million million kilometres away.

Our sun and more than two hundred thousand million other stars are part of our *star system* or *galaxy* — called the milky way. Without a telescope it is possible to see only about 6000 of these stars, even with good eyesight. With a telescope, astronomers can detect other galaxies, further out in space, which they believe to be moving even further from our own galaxy at great speeds. In other words, it seems that the universe is expanding.

Our solar system

One star (our sun), with nine planets in orbit at different distances (see Table 2.1) and with moons (natural satellites) orbiting some planets, is our solar system.

The sun looks small because it is far away, but its diameter (1 393 000 km) is more than a hundred times that of Earth. Like other stars, the sun is bright because it is luminous (produces light). Our moon and the planets do not produce light. They are bright in the night sky because sunlight is reflected by them (bounces off them). This is why Earth, seen from the moon or from a spaceship, appears to shine in the sky.

> **Warning**. It is safe to look at the moon, planets or stars in the night sky, but **never** look at the sun. If you do look at the sun, even through sunglasses, the sunlight will damage your eyes. If you were to look directly at the sun through a telescope or any other optical instrument you would make yourself permanently blind.

Table 2.1   *The planets*

Name of planet	Time for one orbit	Mean distance from sun (millions of km)	Diameter (km)	Number of satellites
Mercury	88 days	58	44 900	0
Venus	225 days	108	12 100	0
Earth	365.25 days	150	12 800	1
Mars	687 days	228	6 800	2
Jupiter	11.9 years	778	143 800	16
Saturn	29.5 years	1 427	120 000	20
Uranus	84 years	2 870	52 300	15
Neptune	165 years	4 500	49 500	8
Pluto	248 years	5 900	3 000	1

Data mainly from North *Mastering Astronomy*.

Day and night and the four seasons

Earth is almost spherical. It is slightly flattened at the north and south poles. Like other planets, Earth moves in two ways. It rotates, like a spinning top, on its own axis (an imaginary line through the north and south poles, see Figure 2.2). It also orbits the sun.

Measuring time. The time Earth takes to orbit the sun is called one year and the time Earth takes to rotate once on its own axis is called one day.

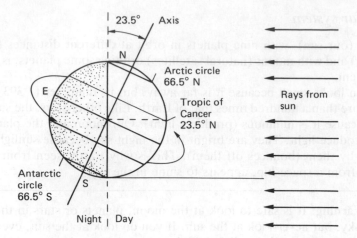

Figure 2.2 Earth's position relative to the sun on 21st June. The direction of radiation from the sun, and the direction of Earth's rotation, are indicated by arrows. N = north pole; E = equator;S = south pole.

But people had to agree to divide each day into 24 hours (symbol h), each hour into 60 minutes (symbol min), and each minute into 60 seconds (symbol s). To be precise, as in athletics, engineering and science, always measure time in seconds.

Day and night. As a result of Earth's rotation (see Figure 2.2), at any moment half is sunlit (daytime) and half is in darkness (night-time).

Four seasons. Because its axis is not at right angles to the sun's rays (see Figure 2.3), each part of Earth experiences four seasons in a year.

June 21st is the longest day in the northern hemisphere and the shortest day in the southern hemisphere because the sun is directly above the Tropic of Cancer. On this day all places north of the Arctic Circle have 24 hours of daylight, while in all places south of the Antarctic Circle the sun does not rise.

December 21st is the shortest day in the northern hemisphere and the longest day in the southern hemisphere because the sun is directly above the Tropic of Capricorn. On this day all places north of the Arctic Circle have 24 hours of darkness and all places south of the Antarctic Circle have 24 hours of daylight.

These dates (June 21st and December 21st) are called the summer and winter *solstices* (Latin: *sol* = sun; *stare* = to stand) because the sun seems to pause on the longest day (before the days start to get shorter) and on the shortest day (before the days start to get longer).

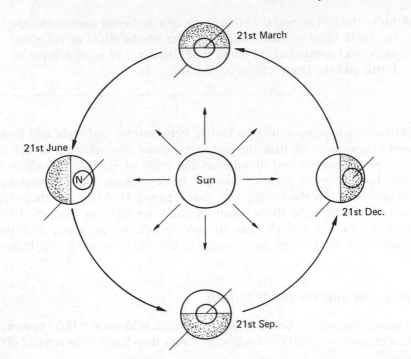

Figure 2.3 Earth viewed from above the north pole at four times during its orbit of the sun. The arctic has 24 hours of daylight on 21st June and 24 hours of darkness on 21st December. The oblique line through the north and south poles represents Earth's axis – an imaginary line.

Half way between the solstices (about March 21st and about September 21st) the sun is directly above the equator. All places on Earth have a 12–hour day and a 12–hour night. Because the day and night are equally long, these dates are called the *equinoxes*.

Investigation 2.1 should help you understand the changes in the position of Earth in relation to the sun (see Figure 2.3) which result in changes in day length and our recognition of four seasons.

Investigation 2.1 *A model to help you visualise Earth orbiting the sun*

You will need a darkened room, a table lamp, a knitting needle, a ball of wool, and a protractor.
Proceed as follows
1 Push the needle through the centre of the ball of wool.
2 Hold the ball of wool with the needle tilted, at about 23° to the vertical, with the needle pointing at one wall of the room.

3 Move the ball around the table lamp, in a darkened room with only the table lamp switched on. Keep the needle tilted at the same angle and pointed at the same wall. The ball of wool represents Earth and the lamp represents the sun.

Differences in temperature on Earth, between day and night and from season to season, result from the fact that the sun produces heat as well as light. The sun warms and illuminates the parts of Earth upon which it shines. In relation to seasonal changes in day length and temperature, there are changes in the life of plants (see Figure 2.4). As a result there are seasonal differences in the amount of food available to animals. Corresponding to seasonal changes in day length, temperature, and the availability of food, there are changes in the life of animals (see Figure 2.5).

Our moon and the lunar month

Our moon, the nearest heavenly body to Earth, is about 400 000 km away. With a diameter of 3500 km it is much smaller than Earth (diameter 12 800 km).

Half the moon's surface is always illuminated by the sun but we see varying amounts of the illuminated part. As a result, the moon appears to change shape (see Figure 2.6).

The moon takes 27 days 8 hours to orbit Earth. The moon and sun are so aligned that there is a new moon every 29.5 days (called one lunar month). As a result, in most calendar months (28–31 days) we see one new moon and one full moon.

The darker areas you see on the moon are the shadows cast by its mountains. Because these shadows always form the same pattern, it is clear that it is always the same side of the moon that faces Earth. For this to be so, the moon must rotate exactly once on its own axis during each orbit of Earth.

Investigation 2.2   *A simple model to help you visualise the moon orbiting earth*

You will need a drawing compass, a pencil, paper and a coin.
Proceed as follows
1 Draw a circle on the paper to represent the moon's orbit of Earth.
2 Move the coin (representing the moon) along the circumference of this circle, keeping the same part of the edge of the coin facing the centre of the circle. As you do this, note that the coin rotates once in each orbit.

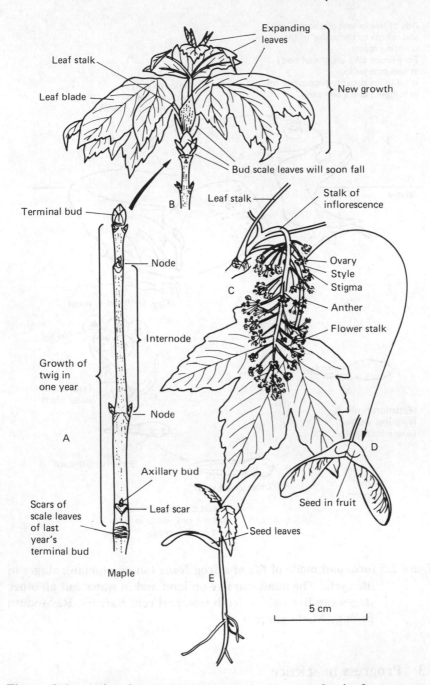

Figure 2.4 a twig of a deciduous tree *Acer,* **a maple, in four seasons: (A) without leaves (in winter); (B) new growth (in spring); (C) inflorescence (in summer); (D) ripe fruit (in autumn); and (E) a seedling (in the following spring). (From Barrass, R.** Modern Biology Made Simple**).**

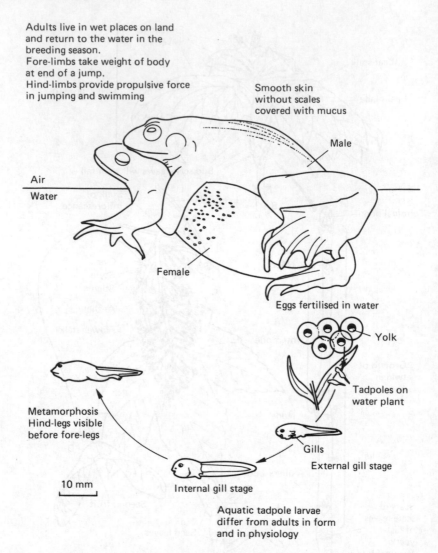

Adults live in wet places on land and return to the water in the breeding season.
Fore-limbs take weight of body at end of a jump.
Hind-limbs provide propulsive force in jumping and swimming

Smooth skin without scales covered with mucus

Male

Air

Water

Female

Eggs fertilised in water

Yolk

Tadpoles on water plant

Gills

External gill stage

Metamorphosis Hind-legs visible before fore-legs

10 mm

Internal gill stage

Aquatic tadpole larvae differ from adults in form and in physiology

Figure 2.5 form and mode of life of a frog *Rana* (an amphibian): stages in life cycle. The adult can live on land and in water but all other stages can live only in fresh water. (From Barrass, R. Modern Biology Made Simple)

2.3   Progress in science

Ptolemy an Egyptian astronomer, over 1800 years ago, was one of the first to record observations and ideas about the universe. In his *Encyclopaedia of Astronomy* he stated that the world was flat and motionless and that the sun, moon and stars moved around it in circles. We know now that this is

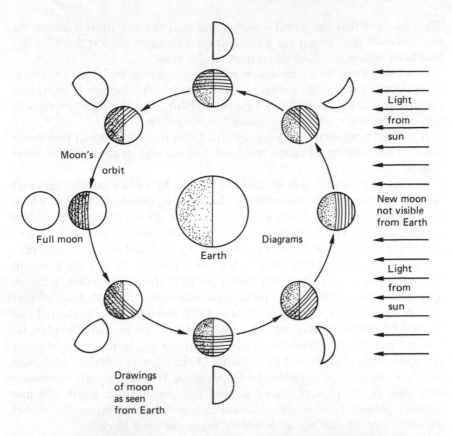

Figure 2.6 Earth and moon viewed from above the north or south pole; and phases of the moon due to changes in the amount of the illuminated half (0 to 100 per cent) that can be seen from Earth. Stippling indicates the half of the moon that is in darkness. Cross-hatching indicates the half of the moon that can never be seen from Earth because it faces away from Earth.

not true, but that does not mean that it was a stupid suggestion. To an observer on Earth, the sun seems to rise in the east, move across the sky and set in the west. Ptolemy's hypothesis was consistent with most of the evidence available to him and provided a possible explanation for most of his observations.

Evidence against Ptolemy's hypothesis was the observation that the planets did not seem to move in circles. So he suggested: (a) that each planet moved not in a circle but in a series of circles or loops, and (b) that the world was not at exactly the centre of the planets' paths.

For hundreds of years Ptolemy was accepted as an authority — most people accepted that his statements were the whole truth. Religious leaders taught that the world had been created at the centre of the heavens.

They insisted that the world was flat and that the sun moved across the heavens each day. This made it difficult for anyone to suggest that Ptolemy had been wrong in his interpretation of the evidence.

However, Ptolemy's conclusions were not accepted by all astronomers. Indeed, about 430 years earlier, Eratosthenes in Alexandria (Egypt) had calculated Earth's circumference (as 40 000 km). But, his proof that Earth was not flat but a sphere was ignored or forgotten.

It was not until about 500 years ago (in 1492) that Christopher Columbus sailed west, hoping to reach the East Indies, and discovered the West Indies.

Nicolas Copernicus, a Pole, had read books by earlier astronomers and he made further careful observations. In 1543 he published a book in which he concluded that: (a) Earth was not flat but a sphere rotating on its own axis once a day; (b) Earth moved in a circle, orbiting the sun once a year; and (c) the sun was at the centre of what we now call our solar system.

You can use a hand lens (magnifying glass) to make nearby objects seem larger than they are (to magnify them). In 1608 Hans Lippershey, a Dutch spectacles maker, described a method of combining lenses to make distant objects seem nearer. Galileo heard about this invention (the telescope) and in 1609 he made a better one. He used his telescope to study the planets.

After months of careful observations, Galileo had further evidence that the apparently complicated movements of the planets could be explained simply — in the way suggested by Copernicus. It was necessary to assume only that all the planets moved around the sun and that Earth was just another planet. Galileo also discovered four moons orbiting the planet Jupiter — proof that not all heavenly bodies orbited Earth.

The evidence collected by Copernicus and Galileo was based on careful observation and calculation. Even though this evidence could have been checked by other astronomers, the conclusions of Copernicus were condemned by political and religious leaders, who placed his book on a list of prohibited books. Galileo at first supported Copernicus but was made to say that he had changed his mind. He had to state in public that he supported the teachings of the church.

The belief that mankind was the purpose of all creation and that the Earth must be at the centre of the universe, had become religious dogma. When something is called dogma it means that it is to be believed as a matter of faith, stated as a truth by a church, political party, or some other authority. It is something which is to be believed to the extent that many people refuse to consider other possibilities. Such fixed beliefs discourage scientific enquiry.

We now know that Earth is not at the centre of the universe. It is not even at the centre of our solar system. It is just one of the planets that orbit the sun. Only Earth's moon orbits Earth.

Furthermore, our sun is not at the centre of the universe – it is just one star near the edge of one galaxy.

③ **Looking at life**

3.1 Differences between living and non-living things

How can you tell if something is alive? Living things include people, pet and farm animals, wild animals and all plants. Non-living things include rocks and all things made by people. Consider how people differ from non-living things. Here are a few differences. You can think of others.

1 People eat and drink. They must do so even if they are not moving about. In contrast, if a car is not used it needs nothing to keep it in good condition. Eating, drinking, breathing, the heart beating and most other activities of the body contribute to *self maintenance*. If any one essential life process stops, soon life will stop. In contrast, a car cannot maintain itself and no damage is caused by switching off the car engine.
2 People can have babies – we say that living things *reproduce their own kind*. In contrast, non-living things cannot reproduce – cars cannot make more cars.
3 People are born, develop and grow through childhood and adolescence to maturity. We call the period from birth to death the *life span*. Non-living things do not grow. Their shape does not change as time passes. If suitably preserved they may remain much the same for hundreds or even thousands of years.
4 If you are much the same today as yesterday this is a result of all the changes going on in your body at all times. Such constancy as there is in a living body is the result of changes. The term scientists use for *a condition of constancy based on change* is homeostasis (Greek: *homoios* = similar; *stasis* = standing) which means remaining steady or a steady state. In contrast to a living thing, if a non-living thing appears the same as time passes this is because there has been little or no change. Things can remain the same in two ways, either by changing constantly (as in life) or by not changing (as in non-living things). All the activities of your body, or those of any other living thing, which contribute to the maintenance of homeostasis, (to the maintenance of the steady state) are called *homeostatic mechanisms*.

The features of all living things, by which we distinguish them from non-living things, are considered further on page 264.

3.2 Different kinds of living things

There are many similarities between all living things. This is why we usually find it easy to tell if something is living or non-living. However, there are many kinds of living things. Scientists call them all organisms. For example, you are an *organism: a living thing*.

Scientists call each kind of organism a species. For example, people are one kind of organism. They all belong to one species.

A *species* is *all organisms of one kind*, different from other kinds of organisms belonging to other species, and able to produce more organisms similar to themselves.

Naming organisms and placing them in groups

Many species of plants and animals have common names, used in everyday language. For example, pea plant and dog are English common names for two species. In other languages different names are used for the same species. For example, the animal called the dog in English is called *le chien* in French and *el perro* in Spanish.

In science we need one name for each kind of organism (for each species) so when one scientist uses a name other scientists will understand which species is meant. The method of naming organisms now accepted by all scientists was devised by Carolus Linnaeus, a Swedish naturalist, in the eighteenth century. This is called the *binomial system* because the name of each species (see Figure 3.1) is in two parts, derived from Latin or Greek words.

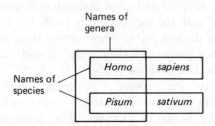

Figure 3.1 species and generic names of organisms

Our own species name is *Homo sapiens*. This is based on two Latin words (*Homo* = man; *sapiens* = wise); we call ourselves wise. Note that in the species name the first word has a capital initial letter and the second word does not. In books both are printed in italics. *Canis familiaris* is the dog; *Pisum sativum* is the garden pea.

An essential part of science, in addition to observing things, recognising them as being of different kinds and giving each kind a name, is to place them in groups according to the things members of each group have in common — by which they can be distinguished from the members of other groups. This grouping of things, according to similarities and differences between them, is called *classification*.

Scientists are not the only people who classify things. People who make and use tools classify them as, for example, hammers and saws. There are many kinds of saw but you have no difficulty in recognising them as saws by the things they have in common — by which you distinguish them from other kinds of tools. It is also possible to classify tools according to the material they are made from (see Table 1.2). This classification is useful to an archaeologist but not to a carpenter.

Living organisms can also be classified in different ways, for different purposes, by different people. For example, farmers classify plants as either crops or weeds. And most people find it helpful to classify plants as either edible or inedible. Scientists classify flowering plants according to the way they grow (as herbs, shrubs or trees) or according to their life span (as annuals, biennials or perennials) and they classify animals according to the kind of food they eat (as herbivores, carnivores or omnivores). These classifications are useful for some purposes but they bring together organisms that have few things in common.

As well as devising the binomial system for naming organisms, Linnaeus suggested a method of placing them into groups that is still used by scientists. He called this a *natural system* because it brought together organisms which had many things in common — by which they could be distinguished from members of other groups.

Linnaeus recognised the species as the basic unit of classification. He grouped all individuals of one kind into the same species, then he grouped similar species into the same genus. He gave all the species in each genus the same generic name. For example, the common frog in Britain is called *Rana temporaria* and the edible frog in France is called *Rana esculenta*. The generic name *Rana* is the name of the genus to which both species belong. The names of these species, their species names, are *Rana temporaria* and *Rana esculenta* (see Figure 3.1).

Table 3.1 *Grouping of species*

Linnaeus grouped	For example, people belong to
similar species in the same genus	genus *Homo*
similar genera in the same family	family Hominidae
similar families in the same order	order Primates
similar orders in the same class	class Mammalia (mammals)
similar classes in the same phylum and similar phyla in the same kingdom	phylum Chordata kingdom Animalia (animals)

We are animals. We are grouped in the phylum Chordata, which includes all animals with backbones and some without backbones. We are mammals — animals with hair whose young are born, (not hatched from eggs with shells). And we are primates — mammals with hands and with eyes that are close together.

Most of the organisms you see every day are either plants or animals. From the drawings (Figure 3.2 and Figure 3.3), and from your own observations, you know that plant and animal life exists in many different shapes. We say that there are many different forms of life.

The plant kingdom

We enjoy looking at plants and many plants are useful to people — they are of economic importance. Moreover, if there were no plants there could be no animals.

Flowering plants. All crop plants, eaten by people and by farm animals, are flowering plants. Cereals are cultivated grasses. These are narrow-leaved plants with dry fruits (called grains) which can be stored from year to year. From broad-leaved plants we obtain vegetables and fruits, spices and drugs, hardwood timbers, cotton and flax, and rubber.

Flowering plants growing where we do not want them are called weeds. These greatly reduce food production from cultivated land but can help to hold the soil (reduce soil erosion) when crops are not growing.

Conifers or cone-bearing plants. From fir trees and other cone-bearing plants we obtain softwood timbers and the pulp used in making paper (see Figure 3.2A and B).

Ferns. Like flowering plants and conifers, the ferns are green plants. They have roots, underground stems called rhizomes, and large leaves called fronds (see Figure 3.2C and D).

Mosses and liverworts. Like ferns, conifers and flowering plants, the mosses and liverworts are green plants that live on land, but they do not have roots. They have delicate thread-like rhizoids that penetrate the soil surface (see Figure 3.2E).

Notes

Algae. The red, brown and green seaweeds that grow on rocky shores and the green scums on fresh water ponds and ditches are classified in the plant kingdom by some scientists (see page 77).

Fungi. Mushrooms and toadstools, moulds that contribute to decay, the mildews, rusts and wilts that cause plant diseases, and ringworm (which is not a worm) and thrush which cause diseases of people, are all fungi. They

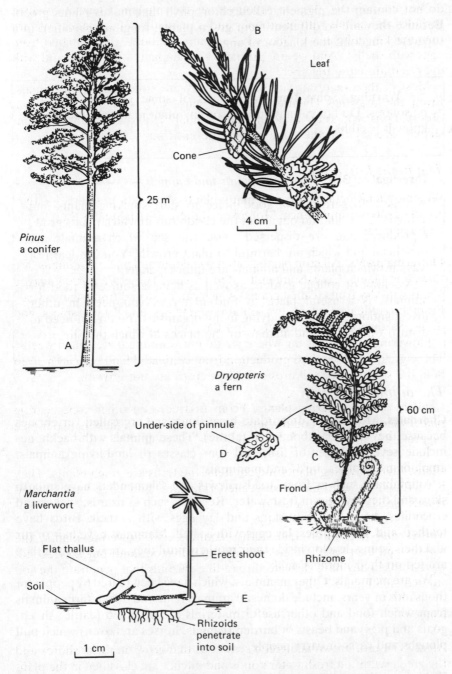

Figure 3.2 plants; (A and B) *Pinus*, **a conifer; (C and D)** *Dryopteris*, **a fern;
and (E)** *Marchantia*, **a liverwort. (From Barrass, R.** Modern
Biology Made Simple)

do not contain the pigment called chlorophyll that makes plants green. Because they are so different from green plants, fungi are classified in a separate kingdom: the kingdom Fungi.

 Warning. Some fungi, and parts of some green plants, are poisonous. Do not eat any fungus, or any plant material, unless you know it is edible.

Investigation 3.1   *Observing plants and animals*

A vegetable garden is a good place to start studying plants and animals. You can find out about the conditions in which plants grow, reproduce, and are dispersed. You can see which animals are beneficial and which are harmful to plant growth. You can learn of ways in which plants and animals are interdependent.

A school or college garden, as well as being a source of food for animals, is a useful place to find materials you need in other investigations, so when studying living organisms you do not need to disturb wild plants and animals or the places in which they live.

The animal kingdom

Chordates. Most kinds of chordate have skulls and are called vertebrates because they have backbones (vertebrae). These animals with backbones include several classes of fish, and four classes of land-living animals: amphibians, reptiles, birds and mammals.

Amphibians such as frogs, toads, newts and salamanders have smooth skins and they lay eggs in fresh water. Reptiles, such as lizards, snakes and crocodiles, have dry scaly skins and lay eggs with a shell. Birds have feathers and, like reptiles, lay eggs with a shell. Mammals have hair or fur and their young develop inside their mothers until they are born, then they are fed on their mothers' milk.

We are mammals. Other mammals, which have been reared by people for thousands of years, include domestic animals (cats and dogs), farm animals from which food and other useful materials are obtained (cattle, sheep, goats and pigs) and beasts of burden (camels, horses and oxen) which pull ploughs and carts or carry people.

Flatworms. Some diseases of farm animals, domestic animals and people are caused by flatworms (flukes and tapeworms, see page 349).

Nematode worms. Although called roundworms, nematodes are cylindrical, tapering at both ends (see Fig. 3.3D). Some infect farm animals,

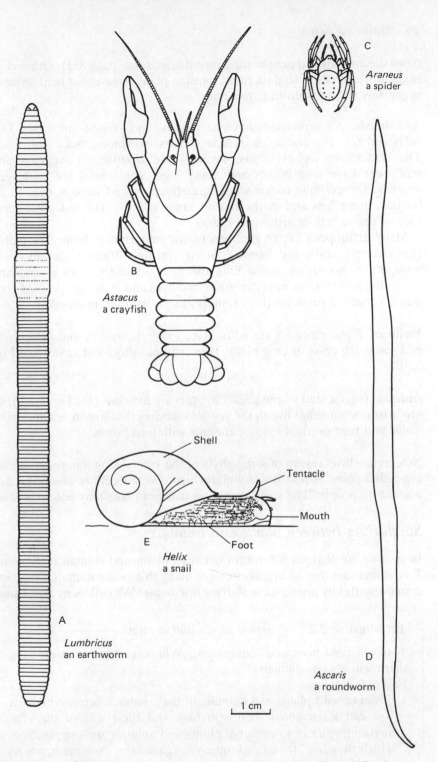

Figure 3.3 animals: (A) *Lumbricus*, **an annelid worm; (B and C)** *Astacus*
and *Aranaeus*, **two arthropods; (D)** *Ascaris*, **a nematode; and**
(E) *Helix*, **a mollusc. (From Barrass, R.** Modern Biology Made
Simple)

domestic animals and people and cause diseases (see page 351). Others live in the soil and either feed on roots, causing plant diseases, or feed on dead organisms and so contribute to decay.

Arthropods. All arthropods have hard skins and jointed limbs (see Fig. 3.3B and C). The *arachnids* include spiders, scorpions, ticks and mites. The *crustaceans* include crabs, shrimps and woodlice. *Centipedes* and *millipedes* have long bodies and many legs. The *insects* are six-legged animals, for example cockroaches, grasshoppers and locusts, lice, fleas, beetles, butterflies and moths and all kinds of flies. This list gives some idea of the variety of arthropod life.

Many arthropods are of great economic importance. Some are edible (for example crabs and locusts). Some transmit diseases (for example mosquitoes, houseflies, tsetse flies and fleas). Some live on people and cause diseases (for example itch mites and lice) and many are pests of crop plants or stored products (for example, locusts and grain weevils).

Molluscs. Some molluscs are edible (for example oysters and octopuses) and some are pests of crop plants (for example slugs and snails: see Fig. 3.3E).

Annelids (segmented worms). Earthworms are annelids (see Fig. 3.3A), so are bristle worms that live in the sea and leeches that live in sea and fresh water and feed on the blood of animals with backbones.

Note In this brief review of some phyla of the animal kingdom some species are called pests. A pest is an animal which, like a weed, is where it is not wanted by people. The same species in another place may not be a pest.

Similarities between plants and animals

In spite of the obvious differences between plants and animals (see Figure 3.4), remember that all organisms have many things in common. Scientists emphasise this by giving them all the same name. We call them organisms.

Investigation 3.2   *Observing plants and animals*

You will need no special equipment. Wherever plants are growing there will also be animals.

Proceed as follows

1 Observe wild plants and animals in their natural surroundings so you can learn about their structure and their way of life. Do anything you can to conserve plants and animals and the places in which they live. If you pick up any organism to examine it, try to leave it as you found it.

2 Wherever possible, study organisms without disturbing them.

Make notes and drawings and take photographs. Keep a record of what you have seen but do not collect specimens.

3 Observe some plants and animals carefully. From their external features, try to place each of them in one of the phyla described in this chapter.

A plant has
(1) A branching body
(2) Green parts in the air and
(3) Non-green parts in the soil
(4) Roots that fix it in one place

An animal
(1) Does not have a branching body
(2) Feeds on other organisms
(3) Has no green parts and
(4) Most animals move from place to place

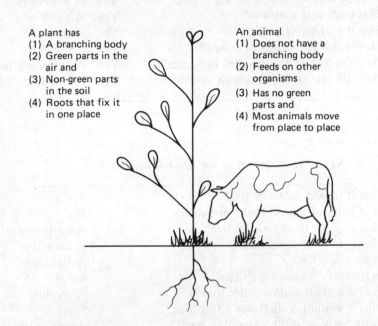

Figure 3.4 differences between plants and animals

Using keys for identifying organisms

Scientists use keys to help identify organisms. With some keys it is possible to find the species names of organisms, but the simple keys provided here are just to allow you to begin to classify the plants and animals you are most likely to see.

Each key (see Tables 3.2 and 3.3) comprises numbered questions. You must look carefully at the organism you are trying to identify, then you will be able to answer yes to either A or B. The words to the right of this answer tell you either the name of the phylum in which the organism is classified or the number of the next question.

Drawing

In preparing records (see page 6) use line drawings, as in this book, to augment your notes — not as an alternative to writing. Work on unlined

Table 3.2   *Key to some phyla of the plant kingdom*

1 A Is it without rhizoids or roots?	Yes	An alga*
B Has it roots or rhizoids?	Yes	See 2
2 A Has it rhizoids?	Yes	See 3
B Has it roots?	Yes	See 4
3 A Is each leaf without a midrib?	Yes	A liverwort
B Has each leaf a midrib?	Yes	A moss
4 A Has it leaf-like fronds?	Yes	A fern
B Has it flowers, fruits or cones?	Yes	See 5
5 A Are the leaves like needles or scales?	Yes	A softwood tree
B Are the leaves flat and with veins?	Yes	A flowering plant

* See also Algae and Fungi (page 22).

Table 3.3   *Key to some phyla of the animal kingdom*

1 A Has it paired limbs or fins?	See 2
B Is it without paired limbs or fins?	See 3
2 A Has it more than four limbs?	An arthropod
B Has it four or fewer limbs or fins?	A chordate
3 A Is its body flat?	A flatworm
B Is its body some other shape?	See 4
4 A Has it a shell and/or a flat foot?	A mollusc
B Has it neither a shell nor a flat foot?	See 5
5 A Is its body of similar ring-like parts?	An annelid
B Is its body unsegmented?	See 6
6 A Has it a cylindrical body with pointed ends?	A nematode
B Is its body some other shape?	See 7
7 A Has it delicate tentacles around a central mouth?	A cnidarian*
B Has it a hard skin with spines?	An echinoderm*

* The phylum Cnidaria includes sea anemones and jellyfishes, which live in the sea. The phylum Echinodermata includes starfishes and sea urchins, which also live in the sea.

A4 paper and use a whole page for each drawing so it is large and clear and you have space for notes. Write notes next to each drawing. Draw labelling lines with a ruler – place your pencil on the point to be labelled and draw a straight line (as in Figure 2.4). Do not draw one labelling line across another. Try to draw the lines so they radiate from the drawing, with lines and labels well spaced away from your drawing.

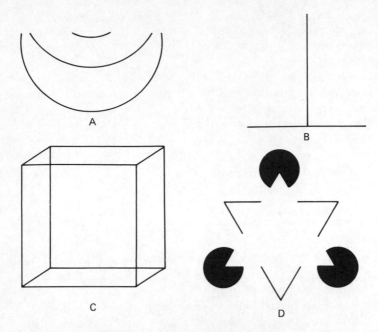

Figure 3.5 optical illusions: (A) arcs of equal radius; (B) horizontal and vertical lines of equal length; (c) a cube which may be interpreted differently at different times; and (D) an illusion of a white triangle. (Redrawn from Barrass, R. Scientists Must Write)

People have always observed, named and classified objects, such as the things they could see in the sky and the different kinds of living organisms. This procedure has provided a firm basis for the further development of many branches of science.

You will find that observing things carefully, classifying them, and giving them names that are accepted by scientists everywhere, is still the starting point in many scientific investigations. However, remember that our interpretation of observations depends on our previous experience. This usually helps us, but we can also be misled (see Figure 3.5).

4 Matter

4.1 The three states of matter

Scientists use the term matter to include all materials. *Matter* is defined as *anything that occupies space*. And a space from which all matter has been removed is called a *vacuum*. Matter exists in three states:

1 A *solid* has both a definite shape and a definite volume, it occupies a fixed amount of space.
2 A *liquid* has a definite volume but not a definite shape, it takes the shape of its container.
3 A *gas* has neither a definite shape nor a definite volume, it takes the shape and occupies the volume of any container.

Measuring matter

Wherever possible, instead of using imprecise words such as few and many, short and long, say how many and how long. By counting and measuring scientists can be precise.

Units of measurement. To measure anything you need a unit of measurement. For example, you could measure distance in paces but even your own paces differ in length. A pace is an example of a non-standard unit of measurement. For all practical purposes standard units are needed so that different people measuring the same thing will make identical measurements.

This is why an international system of units, called SI units (*Système International d'Unités*), is now used by all scientists. Each SI unit is precisely defined and its use contributes to precision in measurement and to international communication (see page 55)

Length. The SI unit for measuring length or distance is the metre (symbol m). In the international system each measurement is recorded as a number followed by a symbol (for example, 1 m = one metre).

Notes

1 A space is left between the number and the symbol.
2 The symbol, unlike many abbreviations, is not followed by a full stop.
3 The same symbol is used in the singular and plural (for example 1 m and 100 m).

In the International System measurements are recorded in whole units, or in multiples or submultiples of units (see Table 4.1). For example 1000 metres = $10 \times 10 \times 10$ metres = 10^3 m is called one kilometre (1 km). One tenth of a metre = 0.1 m is called a decimetre (Latin *decimus* = one tenth). One hundredth of a metre = 0.01 m is called a centimetre (Latin *centum* = hundredth). One thousandth of a metre = 0.001 m is called a millimetre (Latin *mille* = thousandth).

Scientists and engineers write numbers in two ways (for example 1000 or 10^3 and 0.001 or 10^{-3}). Both are correct but the shorter form (as powers of 10) is most convenient when recording very large or very small numbers (see Table 4.1).

The decimal system of counting in multiples of ten (Latin *decem* = ten), must always be used with SI units. This makes the arithmetic simple. The only exception to this rule is when measuring time (60 s = 1 min, 60 min = 1 h). In some countries a comma is used as a decimal point. A comma should not therefore be used to break large numbers into groups of three digits. Numbers up to 9999 are written with no spaces, and with larger numbers gaps are left. (For example, one light year is 9 460 000 000 000 km = 9.46×10^{12} km).

Table 4.1   *Some prefixes and symbols used with SI units*

Number	Factor	Prefix	Symbol
1 000 000	10^6	mega	M
100 000	10^5		
10 000	10^4		
1000	10^3	kilo	k
100	10^2	hecto	h
10	10^1	deca	da
1			
0.1	10^{-1}	deci	d
0.01	10^{-2}	centi	c
0.001	10^{-3}	milli	m
0.0001	10^{-4}		
0.000 01	10^{-5}		
0.000 001	10^{-6}	micro	μ

Note. For further advice on the use of numbers in scientific writing see Barrass, R. *Scientists Must Write*, Chapman & Hall, London.

Investigation 4.1   *How thick is a piece of paper?*

You will need 100 sheets of paper (for example, pages 1 to 200 of this book) and a ruler calibrated in millimetres.

Proceed as follows
1 Measure the thickness of the pile of 100 sheets of paper carefully.
2 To calculate the thickness of one sheet, divide by 100.

Area. The number of unit squares that would just cover a surface, called its area, is always stated in square units. To determine the area of a small rectangle use a ruler marked in millimetres to measure its length and breadth. Then calculate the area in millimetre squares (mm^2) by multiplying these two measurement (see Figure 4.1). You would probably measure the area of land covered by a playing field in metre squares (m^2) and the area of land covered by a city would be measured in kilometre squares (km^2).

--- **Investigation 4.2**   *What is the surface area of a plant leaf?* ---

You will need graph paper marked in millimetre squares, a sharp pencil and a leaf.

Proceed as follows
1 Place the leaf on the graph paper and carefully outline the leaf with your pencil.
2 Count the number of squares inside the outline. This is the area of one surface of the leaf. Double this to give the surface area of the whole leaf (upper + lower surface).

Volume. The amount of space a material occupies, called its volume, is always stated in cubic units (for example mm^3, cm^3 and m^3). The volume of a solid with a regular geometric shape can be determined by direct measurement followed by calculation (see Figure 4.1).

The volume of a liquid is measured by pouring it into a measuring cylinder or jar (see Figure 4.2).

The volume of a solid, even if it does not have a regular shape, can be measured by displacement (see Figure 4.2).

--- **Investigation 4.3**   *Measuring the volume of a solid by displacement* ---

You will need a glass measuring jar, water, and an object that is small enough to fit into the jar and that does not dissolve in water.

Proceed as follows
1 Pour water into the jar. Note that the water curves up the sides, forming a meniscus (see Figure 4.2), so you can see two lines.

2 With the jar on a horizontal surface and your eye level with the meniscus, record the level indicated by the lower line, that is, the base of the meniscus.
3 Submerge the object carefully, then record the new level.
4 Calculate the volume of the object by subtracting your first reading from your second.

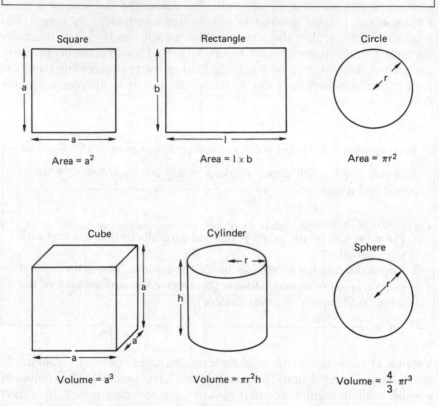

Figure 4.1 calculating area and volume of some regular solids

Mass. The amount of matter any material contains is called its mass. The SI units for measuring mass are the kilogramme (symbol kg) and the gramme (0.001 kg = 1 g).

Investigation 4.4 *Measuring the mass of an object*

You will need an object of unknown mass, a selection of objects of known mass, and an equal arm balance (see Figure 4.3).

Proceed as follows
1 Place the object of unknown mass in one pan of the balance.

2 Add objects of known mass to the other pan. At the point of balance the mass of the object to be measured is equal to the total mass of the objects of known mass.

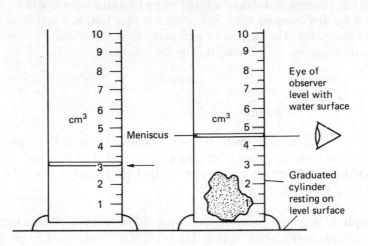

Figure 4.2 a submerged body displaces its own volume of liquid

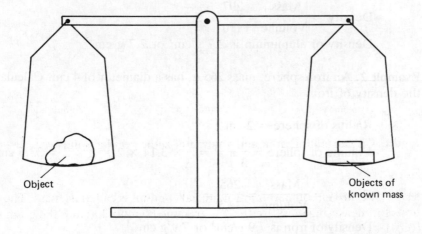

Figure 4.3 using an equal arm balance to determine an object's mass

To find the mass of a known volume of liquid, deduct the mass of the container from the mass of the container plus the liquid.

Density. When you say one material is heavier than another, for example that lead is heavier than aluminium, you mean that if you had a lump of lead and a lump of aluminium of equal volume then the lead would be heavier than the aluminium. You must compare equal volumes of the two materials because you know a large lump of aluminium could be heavier than a smaller lump of lead. Scientists say that lead has a greater density than aluminium. The density of any material is calculated by dividing the mass of a sample of the material by its volume:

$$\text{density} = \frac{\text{mass}}{\text{volume}}$$

Density is defined as *mass per unit volume*. In recording density it is essential to state the units of measurement used. For example, grammes per cubic centimetre (g/cm^3 or $g\ cm^{-3}$) and kilogrammes per cubic metre (kg/m^3 or $kg\ m^{-3}$).

Example 1. A piece of aluminium, mass 297 g, was placed in a measuring cylinder containing 50 cm^3 water. The level of the water rose to the 160 cm^3 mark. From these measurements, calculate the density of aluminium.

Mass of piece of aluminium = 297 g
Volume = 160 − 50 = 110 cm^3

$$\text{Density} = \frac{\text{Mass}}{\text{Volume}} = \frac{297}{110} = 2.7$$

Density of aluminium is 2.7 g/cm^3 or 2.7 $g\ cm^{-3}$

Example 2. An iron sphere, mass 265 g, has a diameter of 4 cm. Calculate the density of iron.

Radius of sphere = 2 cm

$$\text{Volume of sphere} = \frac{4}{3}\pi r^3 = \frac{4}{3} \times 3.14 \times 2 \times 2 \times 2 = 33.5\ cm^3$$

$$\text{Density} = \frac{\text{Mass}}{\text{Volume}} = \frac{265}{33.5} = 7.9$$

Density of iron is 7.9 g/cm^3 or 7.9 $g\ cm^{-3}$

4.2 Physical change

The physical properties of any material can be observed and then described – using adjectives such as hard, soft, strong, brittle, flexible, inflexible, elastic, inelastic, transparent and opaque.

Change of state

It is useful to classify a material as solid, liquid or gas, though you know that water, for example, exists as solid (ice), as liquid water, and as gas (water vapour). The changes of state described by the terms solidify (freeze), liquify (melt), vaporise (evaporate) and condense (see Figure 4.4) are examples of physical changes. All the materials listed in Table 4.2 can exist in other states, depending on temperature.

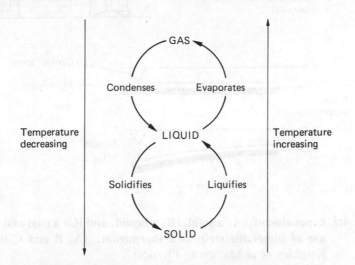

Figure 4.4 changes of state

Expansion and contraction

As well as changes of state, other physical changes associated with changes in temperature are expansion and contraction. As materials are heated they occupy more space — we say they expand (see Figure 4.5). On cooling they occupy less space — they contract.

Table 4.2 *Some solids, liquids and gases*

Solid	Liquid	Gas
Iron	Water	Air
Stone	Oil	Hydrogen
Sand	Mercury	Oxygen

It is easy to demonstrate changes due to expansion and contraction in solids, liquids and gases (see Figure 4.5A, B and C). For a given rise in

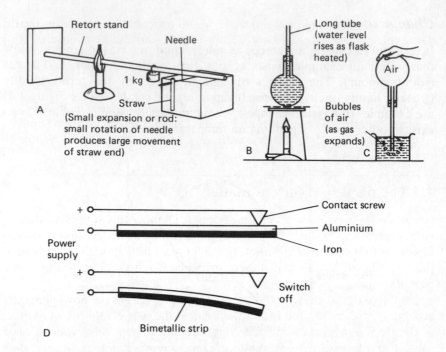

Figure 4.5 expansion of: (A) a solid, (B) a liquid, and (C) a gas; and (D) the use of bimetallic strip as a thermostat. (A, B and C based in Keighley et al Mastering Physics)

temperature gases expand more than liquids, and liquids more than solids.

For a rise in temperature of 100 °C a 1000 mm rod of iron increases by only about 1 mm in length. Although the expansion and contraction of solids on heating and cooling is small, the forces involved (pushes and pulls) are large. This is why structures and machines that have to work at different temperatures must be designed to allow for expansion and contraction. For example, bridges with long spans are made with one end of each span free to move. Similarly, concrete roads are made with gaps between the slabs of concrete to allow for expansion on hot days.

We can also make use of the expansion of solids as their temperature changes. For example, in heat switches (thermostats) two metals with different rates of expansion are used as a bimetallic strip (see Figure 4.5D), to switch an electric current on and off.

Changes in volume as a result of expansion and contraction do not result in any change in mass — there is still the same amount of matter. They do result in changes in density. Remember, density = mass÷volume.

Water, like other substances, contracts and becomes more dense as it cools. But, unlike other substances, its maximum density (at 4 °C) is above its freezing point (0 °C) — so ice floats on water. If it did not, many lakes and seas would be solid ice and there would probably be no life on Earth. The expansion of water as it cools from 4 to 0°C can also be a nuisance. For example, it may cause water pipes to burst.

> **Note.** *After a physical change there is still the same material.* Water is water whether it is solid, liquid or gas. Also, there is still the same mass of material. And *it is usually easy to reverse a physical change* (see Figure 4.4).

4.3 The particle theory of matter

About 2400 years ago a Greek philosopher, Democritus, suggested that all *matter was composed of very small particles* which he called *atoms* (from a Greek word meaning indivisible). For 2000 years little thought was given to this hypothesis, but a great deal of evidence has been collected over the past 400 years.

The hypothesis that matter is composed of atoms is now generally accepted by scientists and is therefore called the particle theory of matter. A theory is a hypothesis that accounts for all known observations on a subject and has gained general acceptance among scientists. As we shall see, atoms are now considered to occur not separately but closely associated with other atoms: for example as molecules or, in some solids, as giant structures.

Investigation 4.5 *Determining the diameter of a molecule*

It is not possible to observe and measure molecules directly as they are too small. In this investigation we assume that a drop of oil spreads on a water surface until it is just one molecule thick.

You will need a fine powder (talcum powder will do), olive oil, a loop of wire, a ruler, a magnifying glass, and a metal tray, at least 50 cm diameter, containing clean water.

Proceed as follows
1 Lightly dust the water surface with the powder.
2 Use the wire loop to pick up a small drop of olive oil about 1 mm diameter. Measure the drop by holding it in front of a ruler and examining it through the magnifying glass.
3 Place the drop on the centre of the water surface where it will spread into a large circular patch. Measure the diameter of the patch.
4 Calculate the thickness of the oil layer (t cm) using the formulae:
area of circle, radius R $= \pi R^2$
volume of disc, thickness t $= \pi R^2 \times t$

volume of oil drop (a sphere, radius r) $= \frac{4}{3} \pi r^3 = \pi R^2 \times t$

Therefore, $t = \dfrac{4 \times \pi \times r^3}{3 \times \pi \times R^2} = \dfrac{4r^3}{3R^2}$

A student performed this investigation and made the following measurements:

Diameter of oil drop = 0.1 cm (r = 0.05 cm)
Diameter of oil patch = 40 cm (R = 20 cm)

If the oil patch is a single layer of oil molecules, the diameter of one oil molecule (t) is:

$$t = \frac{4r^3}{3R^2} = \frac{4\,(0.05)^3}{3\,(20)^2} = \frac{4 \times 0.000\,125}{3 \times 400} = \frac{0.0005}{1200}$$

$$= 0.000\,000\,4 \text{ cm} = 4 \times 10^{-7} \text{ cm}$$

The kinetic theory of matter

The kinetic theory of matter (from the Greek word *kinesis*, meaning movement) states that all matter is composed of molecules that are in constant motion. Although molecules are too small to be seen, there is evidence that they do exist and are constantly moving. One effect considered to be due to the movement of molecules is called Brownian movement because it was observed first by Robert Brown a Scottish botanist in 1827.

Investigation 4.6   *Observing Brownian movement in air*

You will need a microscope, a lamp, a glass cell full of smoke with a glass lid (see Figure 4.6).

Proceed as follows
1 Illuminate the glass cell full of smoke from one side.
2 Observe the smoke through the microscope. You will see smoke particles as small spots of light in constant motion.

This movement of smoke particles in a closed container is evidence that molecules do exist. Presumably, molecules in the air – which you cannot see – are striking the smoke particles from all sides.

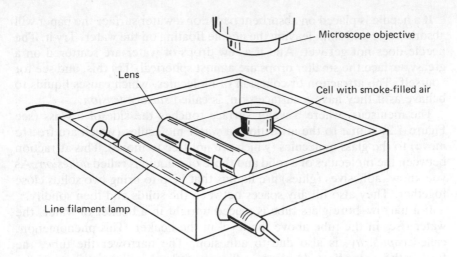

Figure 4.6 apparatus for observing Brownian motion. (From Keighley et al Mastering Physics)

The attraction between molecules. All molecules attract other molecules. *The attraction that causes molecules to hold together* (cohere) is called *cohesion*. The closer they are together the more they attract one another. The size of the spaces between molecules determines the state of a material: solid, liquid or gas.

You may find it helpful to imagine that molecules are very tiny balls. In a solid they are held close together in a regular pattern. They can vibrate but cannot move freely. When a solid is heated they vibrate faster and faster and they take up more space — the solid expands. If heated further the molecules move faster still and may eventually break away from their fixed positions and move freely. This is what happens when a solid liquifies.

The liquid occupies more space than the solid — expansion takes place on melting. The same molecules occupy more space, so the density of a material is less in the liquid than in the solid state. Also, the attraction between molecules is weaker in a liquid than in a solid. This is why a liquid can be poured and takes the shape of its container (see page 31).

The molecules in a liquid are free to move but they attract one another sufficiently to make it difficult for any molecules to escape from the surface. Nevertheless, some molecules do move fast enough to break away. This is evaporation — the liquid turning into a gas. In a gas the molecules are so far apart that there is very little attraction between them. The density of a material is less in the gaseous than in the liquid state. Only a gas is compressible and has neither shape nor volume (see page 31). The moving molecules occupy all the available space and can be kept together only in a container.

If a needle is placed on absorbent paper on a water surface the paper will absorb water and sink, leaving the needle floating on the water. Try it. The needle does not get wet. And if a few drops of water are scattered on a greasy surface the smaller drops are almost spherical. Try this, and see for yourself. The attraction or cohesion of molecules, which causes liquids to behave as if they have a surface skin, is called *surface tension*.

The meniscus, where a water surface touches the side of a glass (see Figure 4.2) is due to the attraction of water molecules (which are free to move) to the glass molecules (which are not free to move). This attraction between the molecules of a solid and those of a liquid is called *adhesion*. As you know, adhesives (glues) are liquids that help to bring two solids close together. They also fill any spaces between the solids and then solidify.

If a narrow–bore glass tube is placed upright in a beaker of water, the water rises in the tube above the level in the beaker. This phenomenon, called *capillarity*, is also due to adhesion. The narrower the tube, the further the water rises. In plants, adhesion and cohesion contribute to the upward movement of water in very narrow tubes (formed from cells) from the root, through the stem, to the leaves.

Porous materials (containing interconnecting small air spaces or pores through which water can permeate) are also called permeable materials. Those through which water cannot pass are called impermeable materials. Some building materials (including bricks, mortar and concrete) are porous. Water can rise through small spaces between the solid particles by capillarity. The foundations of buildings should therefore include a layer of impermeable material (for example, slate or polythene) to stop water rising and making the walls damp. This impermeable layer is called the damp–proof course (see Figure 4.7).

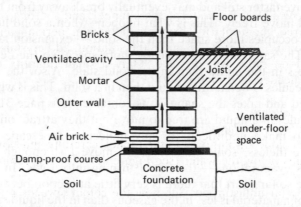

Figure 4.7 two-brick thick cavity wall with impermeable (damp-proof) course. Arrows indicate airflow

Diffusion. When two liquids or two gases are in contact they mix. This is due to *the random movement of molecules*, called *diffusion*, which results in their net movement from regions of higher to regions of lower concentration.

Investigation 4.7 *Diffusion through air*

You will need litmus paper, ammonia, a long glass tube arranged as in Figure 4.8 on a horizontal surface, and eye protection.

Proceed as follows
1 Place squares of wet litmus paper at intervals in the tube.
2 Close both ends of the tube so there is no air movement.
3 Add one drop of ammonia to the cotton wool plug. Note that the litmus papers turn blue as molecules of ammonia diffuse through the tube (see page 149).

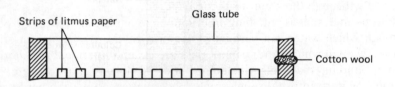

Strips of litmus paper   Glass tube

Cotton wool

Figure 4.8 diffusion of ammonia vapour in air

Investigation 4.8 *Diffusion in a liquid*

You will need a glass jar, water, a glass funnel, and a strong solution of a coloured chemical (for example potassium permanganate or copper sulphate) in water. To make the solution, add crystals to the water until no more will dissolve. Only a small quantity of this solution is needed (see Figure 4.9).

Proceed as follows
1 Two-thirds fill the jar with water.
2 Pour the coloured solution into the funnel, carefully and slowly, so that it runs below the water (see Figure 4.9). Do not remove the funnel.
3 Leave the jar undisturbed. See how the colour spreads gradually through the clear water until all the water is evenly coloured.

Warning. Investigations such as these in which chemicals are used should be performed only in a properly equipped laboratory and only when supervised by an experienced teacher.

Observations in investigations such as these provide evidence that matter is composed of very small particles (molecules) and that in liquids and gases these molecules are free to move.

Do molecules of ammonia diffuse through air more rapidly or more slowly than the molecules of the coloured solution diffuse through water?

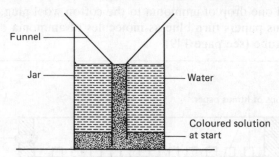

Funnel

Jar

Water

Coloured solution
at start

Figure 4.9 diffusion in a liquid: arrangement of apparatus for Investigation 4.8

You would expect, if molecules are more widely spread in a gas than in a liquid, that diffusion would be more rapid through a gas than through a liquid.

5 | Growth and its measurement

As a crystal grows it gets bigger but its shape does not change. But as a plant or animal grows it does not just get bigger, its shape also changes. Members of a given species have a characteristic body shape at each stage of development and growth.

All the stages in the life of any organism make up its life cycle. For example, each mammal develops in its mother's uterus, grows before and after birth, and may be a parent when mature (see Figure 5.1A).

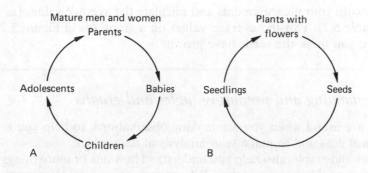

Figure 5.1 life cycles: (A) of people; and (B) of flowering plants

Each flowering plant develops from a seed, grows, flowers, then produces seeds (see Figure 5.1B). Annuals grow from seed, flower, produce seeds in the same year, then die. Biennials grow from seed one year, flower in the next, then die. Perennials grow from seed in one year but do not flower until the second year, or later. Then they may flower and produce seeds for many years.

Some flowering plants are herbs. In the cold or dry season they die back, only their underground parts persist from year to year. Other flowering plants are woody shrubs or trees. Some shed their leaves, usually in a cold or dry season, and are called deciduous plants. Others have green leaves throughout the year and are called evergreens.

5.1 The growth and form of a flowering plant

You cannot tell if seeds are alive just by looking at them. One way to find out is to plant them in suitable soil to see if they grow. Growth is one characteristic by which we distinguish living from non-living things. But what is growth?

Investigation 5.1   *Measuring the volume of bean seeds soaked in water*

You will need ten beans, a measuring cylinder, water, absorbent paper and graph paper.

Proceed as follows
1 Measure the total volume of the beans by displacement (see page 33).
2 Soak the beans in water for an hour. Remove them, dry them on absorbent paper, and then measure their volume again.
3 Return the beans to the water. Repeat the measurements at intervals.
4 Record your measurements and calculate the average values (as in Table 5.1). Plot the average values on a graph (as in Figure 5.2). Do you think the seeds have grown?

Understanding and preparing tables and graphs

Tables are useful when you are making observations, to help you record numerical data and to make your analysis of data easier.

Tables and graphs also help you understand how one or more things vary in relation to changes in another. When you prepare a table or graph, the independent variable — the thing you can choose (the time at which you choose to make your measurements in Investigation 5.1) — must be

Table 5.1   *Change in volume of ten pea seeds* Pisum sativum *when soaked in water.*

Time	Water	Water + peas	Peas	Average
		Volume (cm^3)		
1025	5.0	7.2	2.2	0.22
1125	5.2	8.1	2.9	0.29
1305	6.0	9.4	3.4	0.34
1500	6.0	9.8	3.8	0.38
1700	5.5	9.6	4.1	0.41
1935	5.6	10.0	4.4	0.44
2255	5.2	9.8	4.6	0.46
0935	5.0	9.7	4.7	0.47
2035	5.1	9.8	4.7	0.47

included in the left-hand column of a table (called the stub) or plotted in relation to the horizontal axis of a graph (called the x axis). And the thing over which you have no control (your readings or data, or the results of your analysis of data), called the dependent variable, must be included in the other columns of a table or plotted in relation to the vertical axis of a graph (called the y axis).

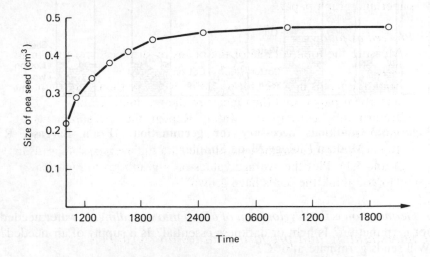

Figure 5.2 increase in volume of seeds of the garden pea *Pisum sativum* **soaked in water. Average values from Table 5.1. (From Barrass, R.** Modern Biology made Simple)

The quantity measured and the units of measurement must be included at the head of each column of a table except the stub and in relation to each axis of a graph. Each table must have a concise heading and each graph must have an appropriate caption or legend.

Investigation 5.2   *Conditions necessary for germination*

You will need five glass tubes or jars, cotton wool, water, cooking oil, black paper to exclude light, a refrigerator at 6 °C and some seeds.

Proceed as follows
1 Arrange the tubes or jars as in Figure 5.3.
2 Look at them each day, or every few days, to see which seeds are germinating. Record your observations.

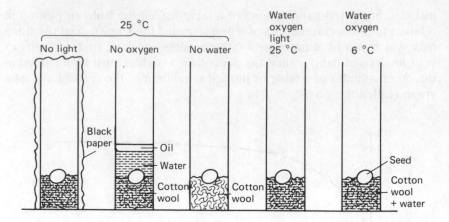

Figure 5.3 conditions necessary for germination. (From Barrass, R. Modern Biology Made Simple)

Germination is *the development of a seed into a seedling.* Is water needed for germination? Is light or darkness essential? Is a supply of air needed? Will seeds germinate at 6 °C?

In moist soil seeds absorb water, increase in volume, and the seed coat may split. A seed that has been soaked in water can be pulled apart and examined (Figure 5.4B). There is a scar where it was attached to its seed stalk in the fruit. The dormant plant, within the seed coat, comprises the young root, two seed leaves and a young shoot.

Investigation 5.3   *Observing germination*

You will need a seed (for example, a pea or broad bean), a glass jar, blotting paper, and water.

Proceed as follows
1 Soak the seed in water overnight.
2 Roll a sheet of wet blotting paper to fit inside the jar, then place the seed between the paper and the glass and add water to 2 cm below the seed.
3 As the young root grows, note the root hairs near the tip. Do the seed-leaves remain in the seed? Compare the leaves formed. When they have expanded fully are they identical?

Investigation 5.4   *The growth of a flowering plant*

You will need a pea or bean seedling, germinating in soil in a pot.

Proceed as follows
1 As the seedling grows into a plant (as in Figure 5.4G), measure the length of the shoot regularly (every few days). At the same times, measure the distance between two axillary buds (the length of an internode, see Figure 5.5).
2 Prepare a table in which you can record the times at which you make measurements and the measurements you make at each time.
3 Plot your measurements on a graph (similar to Fig. 5.2).

What can you conclude about the growth of the plant?

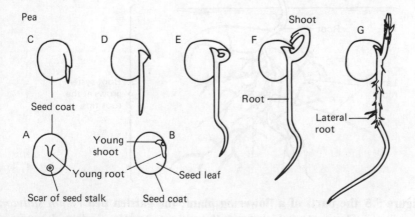

Figure 5.4 structure of seed of garden pea *Pisum sativum* **and germination.
(From Barrass, R.** Modern Biology Made Simple)

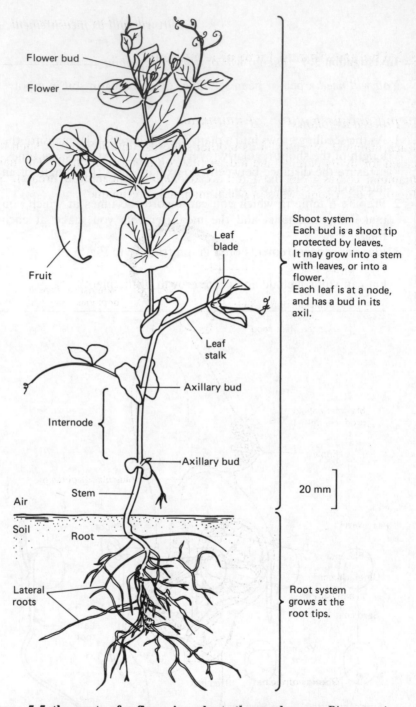

Flower bud

Flower

Fruit

Leaf blade

Leaf stalk

Axillary bud

Internode

Axillary bud

Stem

Air

Soil

Root

Lateral roots

Shoot system
Each bud is a shoot tip protected by leaves.
It may grow into a stem with leaves, or into a flower.
Each leaf is at a node, and has a bud in its axil.

20 mm

Root system grows at the root tips.

Figure 5.5 the parts of a flowering plant, the garden pea *Pisum sativum.* **The roots (which form the root system) and stem, leaves and flowers (which form the shoot system) are all organs. The root system and shoot system are organ systems. The whole plant is an organism.** (Based on Barrass, **R.** Modern Biology Made Simple)

5.2 The growth and form of a mammal

Some unique features of mammals

Each mammal develops in its mother's uterus, until it is born. Then it is cared for by its parents and fed, at first, on its mother's milk. The name mammal comes from the Latin word *mamma* (meaning a breast). See Figure 5.6 for some ways in which mammals are unique.

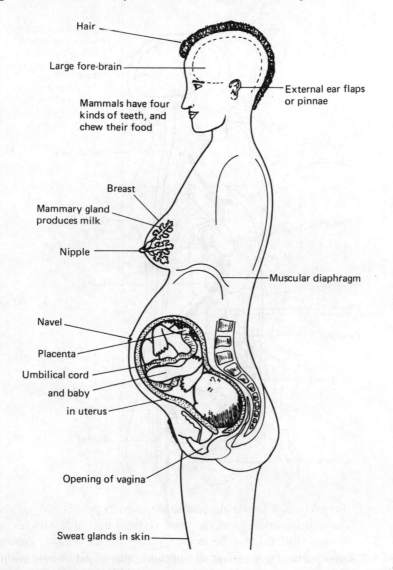

Figure 5.6 characteristics of mammals (From Barrass, R. Human Biology Made Simple)

The parts of a mammal

We recognise different organs (for example, the eyes, brain and kidneys) by position, shape and function. Each organ is part of an organ system. For example, in your own body all the bones, cartilages and joints form your skeletal system. All your skeletal muscles form your muscular system. Your stomach, an organ, is part of your digestive sytem. Your lungs are part of your gas exchange system. Your heart is part of your circulatory system. Your kidneys are part of your urinary system. And your brain is part of your nervous system.

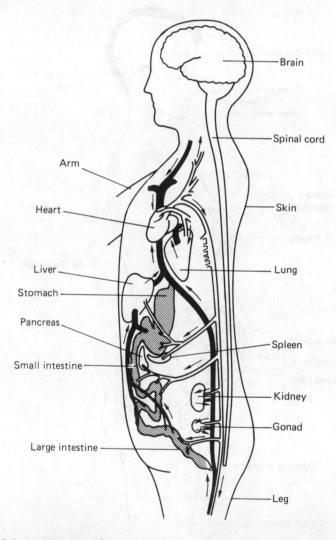

Figure 5.7 some parts of a mammal (simplified). The blood vessels are not labelled but arrows indicate the direction in which blood flows. (From Barrass, R. Human Biology Made Simple)

As in any other organism, the parts of your body are interdependent, that is to say, the normal functioning of your body as a whole depends on the normal functioning of all its parts.

In contrast to a flowering plant, most of the organs of a mammal are not visible externally (see Figure 5.7), but note that the same terms are used: organism (for a plant or animal as a whole), organ system (for the parts of an organism), and organ (for the parts of an organ system).

Investigation 5.5 *Demonstration dissection of a small mammal*

If possible, either in a teaching laboratory or in a museum display, study a dissected small mammal (for example, a mouse, rat or rabbit). Look for the different organ systems.

Do not attempt to dissect dead animals from the wild. They may be dangerous.

Some unique features of people

Most mammals have all the characteristics illustrated in Figure 5.6, by which you can recognise them as mammals, yet each species is unique. For example, we are similar to other mammals yet we differ from them in many ways.

The cerebral hemispheres of the brain of a man or woman are large, compared with those of other mammals. This accounts for our ability to think. We are also unique in our ability to grip things in three ways: in the palm of the hand, between finger and thumb, and between the tips of our fingers and thumb. People are the only animals that can make and use fire, make and use tools, and communicate by speaking and writing.

The growth of a mammal

A mammal, like a flowering plant, changes in many ways as it grows. We use the words babyhood, childhood, adolescence, maturity and old age for the phases of our own growth after birth. Plants and animals grow in different ways but it is not easy to measure growth.

When we want to know how big people are we measure their height or their mass. Yet people of equal height may differ greatly in physique. One person may be physically fit with little or no fat and another may be clearly overweight. If a baby is getting too fat, is it growing more quickly than another baby who is putting on less fat? Would it be better if we could ignore fat in measuring growth? If adding fat is not growth, how can we measure growth?

In an investigation of growth, the tail length of a male mouse was measured from the day it was born until it was fully grown (see Figure 5.8A), as follows:

Age (days)	0	1	3	6	10	17	24	35	51	78
Length of tail (mm)	12	14	18	25	36	45	64	80	88	88

Tables are useful when you are recording data but a graph may be better when you start to analyse data, to help you display and interpret your

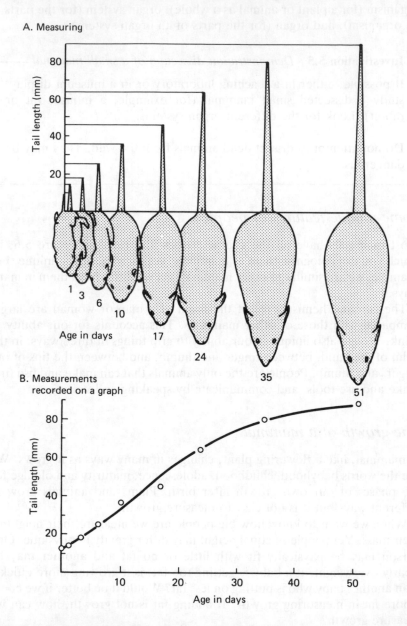

Figure 5.8 the growth of a mouse: (A) measurements of tail length; and (B) the measurements recorded on a graph. (From Barrass, R. Modern Biology Made Simple)

results. For example, in the graph (Figure 5.8B) tail length is plotted against time. From this graph you can see how tail length changes with time. Note as in Figure 5.3 that the variable you can choose (the days on which measurements are made in this investigation) is plotted in relation to the horizontal axis (the x axis) and the variable over which you have no control (tail length in this example) is plotted in relation to the vertical axis (the y axis).

Table 5.2   *International system units of measurement (SI units) used in this book*

Quantity	Unit	Symbol
Length	millimetre (0.001 m)	mm
	centimetre (0.01 m)	cm
	metre	m
	kilometre (1000 m)	km
Area	square centimetre	cm^2
	square metre	m^2
Volume	cubic centimetre	cm^3
	cubic metre	m^3
Mass	gramme (0.001 kg)	g
	kilogramme	kg
Density	kilogramme per cubic metre	kg/m^3
	gramme per cubic centimetre	g/cm^3
Time	second	s
	minute (60 s)	min
	hour	h
	day	d
Speed, velocity	metre per second	m/s
	kilometre per hour	km/h
Frequency	hertz	Hz
Force	newton	N
Pressure	pascal	Pa
Energy, work, quantity of heat	joule	J
Electric current	ampere	A
Power	watt	W
	kilowatt	kW
Electric charge	coulomb	C
Electric potential	volt	V
Electric resistance	ohm	Ω
Thermodynamic temperature	kelvin	K
Temperature	degree Celsius	°C

Note. For further information on SI units and the use of numbers in scientific writing, see Barrass, R. *Scientists Must Write*, Chapman & Hall, London.

6 Forces: pushes and pulls

6.1 Force and movement

Forces are pushes and pulls. A force is needed: (a) to change the shape of an object — as when you compress a steel spring, (b) to start or stop an object moving, and (c) to change the speed or direction of an object that is already moving.

A force must be acting on an object if: (a) its shape changes, or (b) it starts or stops moving, or (c) its speed or direction changes.

Investigation 6.1 *Experiencing some forces*

You will need an elastic band and two spring balances (force meters).

Proceed as follows

1 Hook one finger of each hand into the elastic band and pull with one hand. You will feel that to keep the other hand still it has to pull with equal force but in the opposite direction.
2 To find out if the forces acting in each direction are equal, hook two spring balances in the elastic band in place of your fingers. Pull on one balance to stretch the band but keep the other hand still. Measure the pull on each balance. Are these measurements identical?
3 Hook the elastic band over a fixed object and pull on the band with one hand. This hand is pulling in one direction but where is the second force acting to keep your hand steady? The fixed object must be pulling with an equal force in the opposite direction.
4 Remove the elastic band from the fixed object and again hook one finger of each hand into it. Pull with both hands; then try to exert less force with one hand than with the other. What happens? You will find that this hand will move towards the hand that is exerting the greater force.

From this simple investigation, and from similar everyday experiences, you will understand: (1) that when an object does not move, or moves at a steady speed, this is because the forces acting on it are balanced, and (2) that when the forces acting on an object in opposite directions are not balanced, the object moves (actually accelerates, see page 59) in the direction of the stronger force.

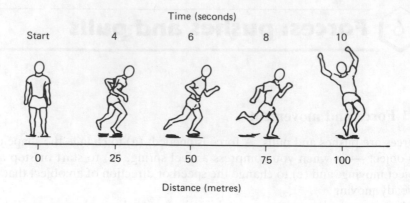

Figure 6.1 average speed = 10 metres per second

We can refer to the speed of an object without reference to the direction of movement. For example:

$$\text{Average speed} = \frac{\text{distance travelled}}{\text{time taken}}$$

Average speed is usually expressed in metres per second or kilometres per hour (see Figure 6.1). If someone runs 200 metres in 25 seconds, the average speed is calculated:

$$\text{Average speed} = \frac{200 \text{ metres}}{25 \text{ seconds}} = 8 \text{ metres per second} = 8 \text{ m/s} = 8 \text{ ms}^{-1}$$

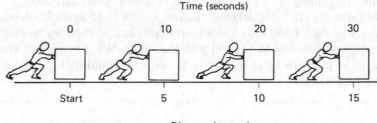

Figure 6.2 pushing a box at a steady speed

When you push a heavy box you can feel the force needed to make it move. All forces are *vector quantities*: they *act in a particular direction*. If you push a box at a steady speed it will move in the same direction as the force applied (see Figure 6.2).

An object's *velocity* is its *speed in a particular direction*.

$$\text{Velocity} = \frac{\text{distance travelled in a particular direction}}{\text{time taken}}$$

When you stop pushing a box it stops moving. For many centuries, as a result of observations such as this, people thought a constant force was needed to produce a constant velocity and that if no force was acting there would be no movement. They were wrong.

In the seventeenth century Galileo showed that a force is needed to start an object moving but once moving it will continue at the same speed and in the same direction (that is, at the same velocity) unless it is affected by some other force.

The reason a heavy box stops moving when you stop pushing it is that there is another force acting between the bottom of the box and the floor. This force, called *friction*, acts in the opposite direction to the movement.

The faster a car is moving the greater the force (friction) needed to stop it. So, even if maximum force is applied to the brakes, the faster a car is moving the longer it takes to stop.

If you are pushing a box you can minimise friction by putting the box on wheels or rollers. Then a constant force will produce not a constant velocity but an increasing velocity. *The rate of increase in velocity* is called *acceleration* (see Figure 6.3).

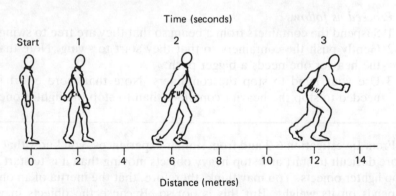

Figure 6.3 accelerating from a standing start

$$\text{Acceleration} = \frac{\text{change in velocity}}{\text{time taken for change}}$$

If velocity is measured in metres per second, acceleration is expressed in metres per second, per second (that is, velocity per second):

$$\text{m/s per second} = (\text{m/s})/\text{s} = \text{m/s}^2 = \text{ms}^{-2}$$

(These equals signs indicate different ways of expressing the same thing.)

When you are in a vehicle that suddenly stops, you feel as if your body is thrown forward. Like all moving objects, your body continues to travel at the same speed and in the same direction (at the same velocity) until another force acts upon it. This *resistance to change in motion* is called *inertia*. To stop yourself being thrown forward you can hold some object fixed in the vehicle so that your velocity decreases with that of the vehicle.

When a car involved in an accident stops suddenly, the driver and all the passengers keep moving at the speed the car was travelling at immediately before the accident. This is a major cause of injury, when people strike the front of the car or the people in the front seats. Seat belts help prevent such injuries.

Objects that are not moving also possess inertia. That is why it is more difficult to start an object moving than to keep it moving.

Investigation 6.2   *Inertia of stationary and moving objects*

You will need two identical containers, one filled with sand or soil and the other empty, and some thick string or rope.

Proceed as follows
1 Suspend the containers from a beam so that they are free to swing.
2 Gently push the containers so that they start to swing. Note that the heavier one needs a bigger push.
3 Use your hand to stop the containers. Note that more effort is needed to stop the heavier container than to stop the lighter one.

From Investigation 6.2 and from similar experiences, you know that it is more difficult to start and stop heavy objects moving than it is to start and stop lighter objects. You may think, therefore, that the inertia of an object depends on its weight. But this is not so. Because the objects in your investigation were suspended, their weights (the pull of gravity on the objects) were balanced by the upward pull of the strings.

The inertia of an object remains the same in space (for example, in a space vehicle), even though it weighs less or seems weightless. In fact the inertia of an object depends only on its mass, so we can use the word mass instead of the word inertia. The greater an object's mass, the greater its inertia.

Newton's laws of motion

Isaac Newton, an English physicist born in the year of Galileo's death, continued the study of forces begun by Galileo. In 1687 Newton summarised in three laws of motion all that had been observed about the movement of objects.

Newton's first law is that *every object remains at rest or moves in a straight line at a constant velocity unless acted upon by a force*. On Earth air always offers some resistance to movement, but in space, where there is no air and so no friction, an object can move forever once it has been started moving by a force. This is why space vehicles and satellites, once they have been put into space, do not need fuel to keep them moving.

Newton's second law is that *when an object is acted upon by a force the acceleration produced is proportional to the force*. The acceleration produced by a force depends on the mass of the object being moved. An object with greater mass requires more force to produce a particular acceleration than would an object of smaller mass.

force = mass × acceleration

The SI unit for force is the newton (symbol N). *One newton is the force needed to accelerate a mass of one kilogramme one metre per second, per second.* $1 \text{ N} = 1 \text{ kg/s}^2$

Newton's third law is that *for every action there is an equal and opposite reaction*. That is to say, forces occur in pairs. When one object exerts a force on a second object, the second object exerts an equal but opposite force on the first. Note that there are two objects and two forces. When you pull an elastic band with a certain force, the elastic band exerts an equal force on your finger — acting in the opposite direction. This is the force that you can feel. And when you walk, run or jump, pushing backwards on the ground with your foot, the ground pushes forward with equal force, due to friction, and you move forward. If the ground is not firm, as when you cross powdery dry sand on foot, friction is reduced and it is difficult to move forward.

6.2 The force of gravity

The force that pulls objects towards Earth is called gravity. *The weight of an object is a measure of the force of gravity acting on the object.*

Gravity acts between all masses, but the magnitude of the force weakens as the masses get further apart. The weight of an object on Earth, therefore, depends on its mass and on its distance from Earth's centre.

Weight = mass × force of gravity

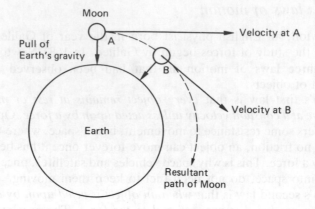

Figure 6.4 the moon orbiting Earth

Because Earth is not a perfect sphere (see page 11), an object is nearer to the centre if it is at either pole that it would be at any other point on the surface. As a result, an object would weigh about 0.5 per cent more at either pole than at the equator. For the same reason, an object would weigh more at sea level than on a mountain top. Differences in an object's weight on Earth are not great, but changes in weight would be much more noticeable if the object were taken into space.

It is the gravitational pull of Earth that keeps the moon — and artificial satellites — in orbit. And it is the gravitational pull of the sun that keeps Earth — and the other planets — in orbit. *Gravity acts towards the centre of every astronomical body* (see Figure 6.4).

Because of its smaller mass, the gravitational pull of the moon is much weaker than that of Earth. But the moon's gravity does affect Earth. Tides are due to the gravitational pull of the moon and sun on the oceans. The sun has less effect than the moon because although the sun has a much greater mass than the moon it is also much further away from Earth. The highest and lowest tides (called spring tides) occur every 14 days, at new moon and at full moon when Earth, the moon, and the sun, are in line (see Figure 6.5).

Measuring weight

The increase in length of a spring is proportional to the force acting on it, so the stretching of a spring can be used to measure the size of a force (see Figure 6.6). A spring balance can therefore be used to measure the weight of an object: that is to say, the force of gravity acting on the object. Although called a spring balance, it is really a force meter or newton meter (not a balance, see Figure 4.3). It should therefore be calibrated in newtons (the SI units for force), not in mass units.

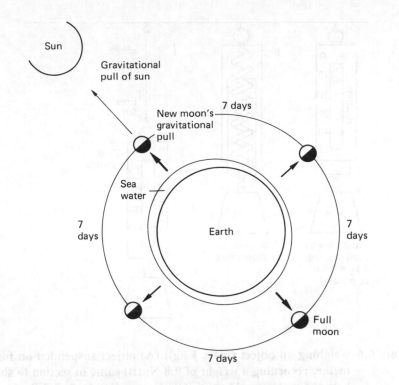

Figure 6.5 Earth viewed from above the north pole. Highest and lowest tides (spring tides) when gravitational pull of sun added to that of moon (at new moon); 7 and 21 days later (at neap tides), the tide does not come in as far and does not go out as far. The tidal bulge is on the side of Earth nearest to the moon (and at the same time on the opposite side of Earth, furthest from the moon). There are two high tides and two low tides every 24 h 50 min

In practice, many spring balances are calibrated in mass units (grammes or kilogrammes). The gravitational force that acts on a mass of one kilogramme, at sea level, is about 9.8 newtons. So for most practical purposes, to convert kilogramme readings to newtons it is accurate enough to multiply by ten.

In everyday (non-scientific) language, many people say weight when they should say mass. For example, people who say they want to lose weight could do so by going to the moon. A boy with a mass of 60 kg on Earth (weight about 600 N) would weigh only about 100 N on the moon, but his mass would still be 60 kg.

The mass of an object does not change unless the object changes. A hammer taken into space will float around in a space vehicle but it is still a

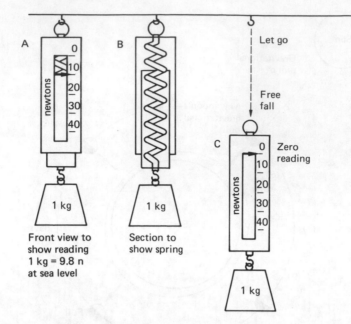

Figure 6.6 weighing an object (mass 1 kg): (A) object suspended on force meter, recording a weight of 9.8 N; (B) same in section to show how the meter works; and (C) zero reading in free fall

hammer. Although it may seem weightless, the hammer still has all the properties it had on Earth. It contains the same materials and can be used to drive in nails.

To avoid confusion, always refer to the quantity of material as mass. Use the term weight only when you are dealing with gravitational force and always measure weight in newtons. When you need to measure mass accurately, use an equal arm balance (see Figure 4.3).

Investigation 6.3   *Free fall*

You will need only a brick and a half brick.

Proceed as follows
To test the hypothesis that if two objects are released together from the same height they will hit the ground at the same time, hold a brick and a half brick at the same height above the ground and then release them together. Does one hit the ground before the other?

Galileo was probably the first to try this experiment. The force acting on the brick is twice the force acting on the half brick. But the brick has twice the inertia (mass) of the half brick. So their acceleration is identical and they hit the ground at the same time.

Objects with a large surface area but very little mass (for example, a parachute and a feather) fall to Earth slowly because of air resistance. Similarly, small objects such a dust particles, pollen grains, and the spores of fungi, have a much larger surface area in proportion to their volume than larger objects. Because of air resistance they fall to Earth slowly. But note that all objects in a vacuum fall to Earth with an acceleration of about 9.8 metres per second per second.

One kilogramme suspended on a spring balance (see Figure 6.6A) weighs about 9.8 newtons. But the reading is zero when both are in free fall (Figure 6.6C). Note that a satellite, or an astronaut orbiting Earth, is in free fall (see Figure 6.4). That is why the astronaut has a feeling of weightlessness. In fact the satellite and astronaut both have weight — the pull of gravity provides the centripetal force needed for circular motion (keeping them in orbit).

The centre of gravity and stability

The force of gravity acts on every part of an object. If you rest one ruler flat on the edge of another (as in Figure 6.7A), it will balance only at its centre — like a see-saw. It is as if the whole weight of the ruler were acting through one point. We call this point of balance the *centre of gravity*. In a see-saw the centre of gravity is immediately above the pivot, so the whole weight of the beam acts through the pivot.

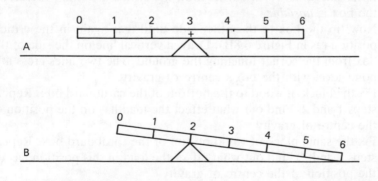

Figure 6.7 balancing forces

The higher the centre of gravity of an object, the more likely is it to fall over when tilted. Remember this when you load a vehicle or use a filing cabinet. Is it best to carry a heavy load in the boot of a car or on a roof rack? Which drawer of a filing cabinet should be loaded first? What could happen if you pulled out the top drawer of a filing cabinet full of papers, when the other two drawers were empty?

Investigation 6.4 *Balancing forces*

You will need two identical rulers and a sharp edge to use as a pivot.

Proceed as follows
1 Arrange one ruler on the pivot as in Figure 6.7B. If the whole weight of this ruler acts through its centre of gravity, where would you expect to have to place the second ruler (across the first) to balance the first ruler on the pivot?
2 Place the second ruler across the first so that the first ruler does balance on the pivot. What do you conclude?

Investigation 6.5 *Finding the centre of gravity of an object*

You will need a cardboard box, a piece of wood that fits into the bottom of the box, a set square, and a pencil.

Proceed as follows
1 Tip the empty box to one side until it is just at its point of balance (as in Figure 6.8A). Draw a vertical line on the side of the box from the corner touching the ground. If you did not tip the box as far as this critical point, the box would fall back on its base when released. Up to the *critical position* the box is said to be *stable*. If you tipped the box further than the critical position, it would fall on to its side when released. Beyond the critical position, therefore, the box is *unstable*.
2 Now tip the box to the other side until it is again in the critical position (as in Figure 6.8B). Draw a vertical line on the side of the box from the corner touching the ground. The two lines cross at a point level with the box's centre of gravity.
3 Fix the block of wood to the bottom of the cardboard box. Repeat steps 1 and 2. Find out what effect the load has on the position of the centre of gravity.
4 Fix the same block of wood on top of the cardboard box. Repeat steps 1 and 2. Find out what effect the load in this position has on the position of the centre of gravity.

6.3 Pressure

A drawing pin is shaped so it is easy to press, to drive in the point. A large force can be applied over a very small area — the pin point. In contrast,

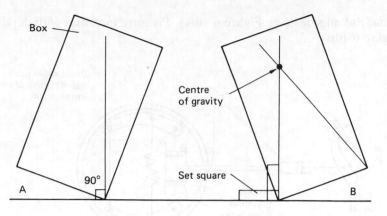

Figure 6.8 finding the centre of gravity

someone wearing skis would not sink as far into snow as someone without them. With skis the same force is applied over a larger area. *Pressure is the force (or thrust) acting per unit area.*

$$\text{Pressure} = \frac{\text{Force}}{\text{Area}}$$

Force is always measured in newtons and area is measured in square metres. The SI unit for pressure is the pascal (symbol Pa). One *pascal* is defined as *a force of one newton acting on one square metre.*

$$1 \text{ Pa} = 1 \text{ N/m}^2$$

Find out your own weight in newtons (1 kg = 9.8 N). Find the area of the sole of one of your shoes (see page 33). How much pressure do you put on the ground when you stand on one leg, and when you stand on both legs? What would be the effect of wearing skis, if their area against the ground is 0.3 m²?

When a water pistol is full of water, pressure at one end forces out a jet of water at the other end. This is because: (a) liquids are virtually incompressible, and (b) pressure upon the liquid at one place is transmitted equally to all other parts. These two facts are used in the design and operation of the hydraulic brakes of a car (see Figure 6.9). A force on the brake pedal results in the brakes being applied some distance away on the wheels.

As you can see when a liquid escapes through holes of equal diameter in the side of a cylinder (as in Figure 6.10A and B), liquids exert pressure sideways as well as downwards. At a given depth, pressure is exerted equally at all sides (see Figure 6.10A). Pressure increases with depth (see Figure 6.10B).

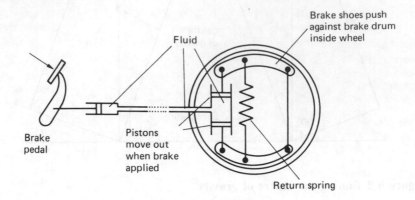

Figure 6.9 how hydraulic brake works. (Based on Keighley et al Mastering Physics)

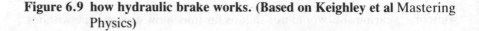
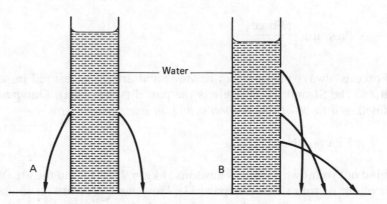

Figure 6.10 pressure exerted by a liquid

A solid exerts pressure downwards only. All *fluids* (liquids and gases) *exert pressure equally in all directions*. Because the pressure in a column of liquid increases with depth, the force acting upwards on the lower part of a submerged body is greater than the force acting downwards on the upper part. That is to say, there is *a net upward force* called *buoyancy* or *upthrust*. One result is that an object weighs less when immersed in liquid than when suspended in air. You can see this yourself if you suspend an object from a spring balance (force meter) first in air and then with the object suspended in water.

If the downward force due to gravity acting on a submerged object (the weight of the object) is greater than the upward force (the weight of the fluid displaced) the object sinks. If the downward force due to gravity (the weight of the object) is balanced by an equal upward force (the weight of the fluid displaced) the object floats. The *law of flotation: a body floating in a fluid displaces its own weight of the fluid*, was discovered by the Greek philosopher Archimedes just over 2200 years ago.

Investigation 6.6   *Compressing air*

You will need a bicycle pump (or a similar air pump).

Proceed as follows
1 Pull out the plunger of the pump, cover the opening with your finger to trap some air in the pump, and then press the plunger until you feel the air pressing against your finger. Note that air is compressible: the air trapped in the pump can be squeezed into a smaller space.
2 Keep pressing and then release the plunger. What happens to the compressed air when you stop squeezing it? (See page 31).
3 Record your observations.

A valve is a device that allows a fluid (liquid or gas) to pass in one direction only. As you pull the plunger of a bicycle pump, air from outside enters the pump (See Figure 6.11A). The valve on the bicycle tyre prevents air from the tyre entering the pump. When you push the plunger, air

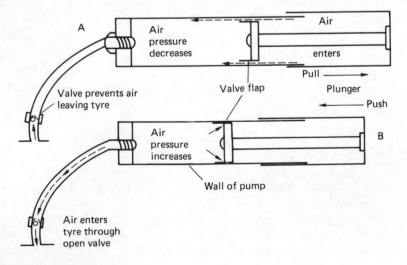

Figure 6.11 an air pump

trapped in the pump is compressed, holding the pump valve flap against the wall and so closing this valve (see Figure 6.11B). The compressed air is forced through the tyre valve into the tyre.

The air pressure in the tyre increases as more and more air is squeezed into the available space. The tyre becomes hard. Note, as with hydraulic brakes (see page 67), a force applied at one place has an effect some distance away.

The pressure of outside air, which causes, for example, water to flow into a teat pipette as the pressure inside decreases, or air into an air pump, (see Fig. 6.11A), is called atmospheric pressure. Because of the pull of Earth's gravity, the molecules in air are more tightly packed near the ground than higher in the atmosphere. Above 40 km there is very little air. Imagine an ocean of air, with the air near the ground squeezed and made more dense by the weight of air above. We are not bowed down by this great pressure (over 100 000 N/m^2) because gases exert pressure equally in all directions.

Because the density of air decreases with altitude, the pressure on the upper part of a balloon is much less than that on the lower part. So a balloon rises in still air if the air it displaces weighs more than the balloon. The balloon remains suspended at the height at which the weight of air displaced is equal to that of the balloon. At this height, the downward force due to gravity is balanced by the upward force (buoyancy or upthrust).

7 A closer look at life

Many scientific instruments extend our ability to observe. In this way they make new observations possible. For example, with binoculars or a telescope you can see objects or details some distance away that you would not be able to see unaided.

Holding something close to your eyes, if you have good eyesight you can see details that are only 0.1 mm apart. When things are closer together than this (the limit of our eyes' resolving power) we do not see them as separate objects.

Investigation 7.1 *Using a magnifying glass or hand lens*

When you observe small organisms, or parts of larger organisms, a hand lens helps you to see more detail than you could see without magnification.

You will need a lens that magnifies things ten times ($\times$ 10) or, preferably, an instrument with $\times$ 10 and $\times$ 20 lenses.

Proceed as follows
(1) Keep both eyes open.
(2) Hold the lens still (about 8 cm from one eye).
(3) Move the object you are examining, in a good light, until it is in focus.

The first microscope was made by Leeuwenhoek in Holland in the seventeenth century. It had only one lens (a simple miroscope) but with it Leeuwenhoek discovered living things that nobody had seen before.

Small organisms, that can be seen only with a microscope, are called micro-organisms or microbes (Greek: *mikros* = small; *bios* = life).

A compound microscope has more than one lens. It gives greater magnification than a simple microscope. The highest magnification used with a compound microscope is normally $\times$ 1000, enabling us to see things that are only 0.0001 mm (= 0.1 μm) apart.

7.1 Unicellular organisms

Two unicellular protists

Amoeba, a microscopic organism that moves over mud at the bottom of freshwater pools and ditches, is just visible without a microscope. With a microscope you can see its structure. The parts of its body are called the *nucleus* and the *cytoplasm* (see Figure 7.2) and the outermost part of the cytoplasm is called the cell *surface membrane*. Such a body is called a *cell*. *Amoeba* is therefore described as a one-celled or *unicellular organism*.

In a drop of pond water you are likely to see other unicellular organisms. Some of these, like *Chlamydomonas* (see Figure 7.2), have one or more chloroplasts in their cytoplasm. Each *chloroplast* contains a green pigment called *chlorophyll*. This is the pigment that makes plants appear green. As we shall see, without this pigment plant life would be impossible, and without plants there could be no animals. See protists, plants and animals in Table 7.1.

Many kinds of unicellular organisms live in fresh water, many others live in the sea and many others live in the bodies of animals. For example, the diseases malaria and sleeping sickness are caused by unicellular protists injected into people by blood-sucking insects when they feed.

Investigation 7.2   *Looking for unicellular organisms in water*

You will need fresh water from a ditch or pond*, a needle, some absorbent cotton wool, a clean microscope slide and cover slip, and a microscope. Avoid rivers and streams where there may be rats. Cover any cuts with a waterproof bandage **before** the sample is taken.

Proceed as follows
1 Place a drop of pond water near the centre of the microscope slide. Add a few threads of absorbent cotton wool.
2 Pick up the cover slip by two opposite edges. Place a third edge so that it touches one side of the drop of water (see Figure 7.1). Hold the needle under the cover slip and lower it slowly until its whole surface is in contact with the water.
3 Examine your preparation using a microscope.

---

* or rain water in which grass clippings have stood for a few days

A unicellular fungus

Investigation 7.3   *Looking at yeast cells*

Yeasts occur naturally on the surface of fruit. They are also cultured commercially for use in brewing and baking (see Figure 7.2).

You will need some live yeast, glucose solution (5 per cent by volume), a beaker, a clean microscope slide and cover slip, a needle and a microscope.

Proceed as follows
1 Make a suspension of yeast (10 per cent yeast by volume) in the glucose solution, in the beaker.
2 Allow this suspension to stand in a warm place overnight (at about body heat).
3 Examine a drop of this suspension on a microscope slide.

Note. Yeast, like *Chlamydomonas*, has a cell wall outside its surface membrane. Unlike *Chlamydomonas*, yeast has no chloroplasts. It has a nucleus in its cytoplasm (see Figure 7.2) but you would need special dyes if you wanted to stain this in your preparation.

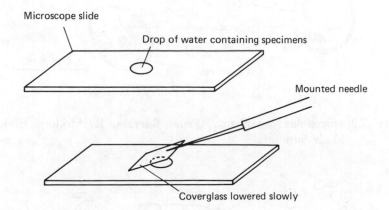

Microscope slide

Drop of water containing specimens

Mounted needle

Coverglass lowered slowly

Figure 7.1 mounting material for microscopic examination. (From Barrass, R. Modern Biology Made Simple)

Bacteria: even smaller unicellular organisms

Bacteria are present everywhere on Earth: in air, water and soil; upon and within the bodies of other organisms. Some are harmful — causing diseases — but most are beneficial

Leeuwenhoek, with his simple microscope, was the first to see bacteria. They are very small and it is not possible to see details of their structure even with a compound microscope. Some are spheres (cocci), some are rods (bacilli), some are spirals (spirilla) and some are curved (vibrios) (see Figure 7.3A).

Our knowledge of the detailed structure of bacteria is based on studies using electron microscopes. A magnification of × 100 000 is commonly used, resolving structures that are only 0.000 001 mm apart. From such studies it is clear that each bacterium has a cell wall, a cell surface

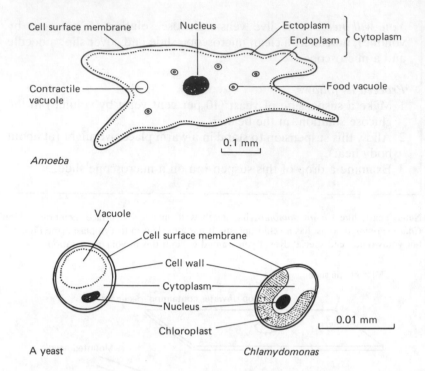

Figure 7.2 unicellular organisms. (From Barrass, R. Modern Biology Made Simple)

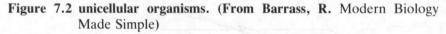

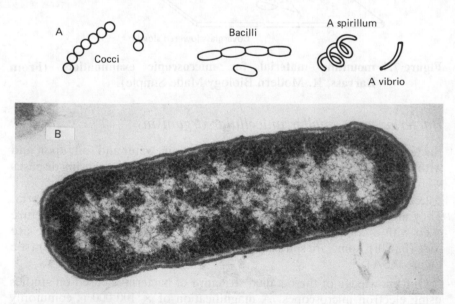

Figure 7.3 bacteria: (A) different kinds of bacteria; and (B) ultrastructure of a bacillus. Electronmicrograph prepared in the School of Biology, Sunderland Polytechnic

membrane next to the cell wall, and a nuclear area (the lighter part of the cell in Figure 7.3B). There is no membrane between the nuclear area and the cytoplasm. In this and other ways the cells of bacteria differ from those of unicellular protists and fungi (see Figure 7.2).

7.2   Multicellular organisms

The invention of the microscope enabled biologists not only to observe very small organisms for the first time but also to investigate details of the structure of larger organisms. If you have a microscope, try the following investigations.

Investigation 7.4   *Examining the pulp of a tomato fruit*

You will need a ripe tomato, a microscope slide and cover slip, and a microscope.

Proceed as follows
1 Place a drop of the soft pulp from immediately below the skin of the tomato on the microscope slide and mix with a drop of distilled water (if not available use tap water).
2 Lower the cover slip on to this fluid (see Figure 7.1).
3 Examine your preparation under low power then under high power. Note the cells floating in the fluid and the red pigment in spherical structures, called plastids, in each cell.

Investigation 7.5   *Examining the epidermis (skin) of an onion leaf*

You will need an onion, iodine solution, a microscope slide and cover slip, and a microscope.

Proceed as follows
1 Each scale of the bulb is a swollen leaf base. Remove one of these leaves and break it so you can peel away the delicate skin.
2 Place a small portion of this skin (the epidermis) on the microscope slide in a drop of iodine solution.
3 Add a cover slip (as in Figure 7.1) and examine your preparation using a microscope. Note that the skin is composed of cells (see Figure 7.4A) and iodine is absorbed by some structures in these cells, enabling you to see these structures.

Investigation 7.6   *Epithelial cells from the skin of a mammal*

You will need a fresh eye of a mammal from a butcher's shop (this can be refrigerated for up to four days), a microscope slide and cover slip, a needle, 1% methylene blue and a microscope.

Proceed as follows

1 Press the slide gently on to the transparent skin (cornea, see Figure 17.13 of the eye. Some cells from the surface of the cornea will stick to the slide.
2 To stain them, flood the slide with three drops of methylene blue.
3 Add the cover slip (see Figure 7.1) and examine your preparation under low power then under high power.
4 Record your observations. Methylene blue stains cytoplasm pale blue and nuclei dark blue. How do these animal cells differ from the plant cells studied in Investigations 7.4 and 7.5? Also note any similarities.
5 Wash hands thoroughly and scrub bench with disinfectant.

Note. In places where a mammal's skin is moist, for example the cornea of the eye, the outer cells of the epidermis (see Figure 7.4B) are alive and have nuclei.

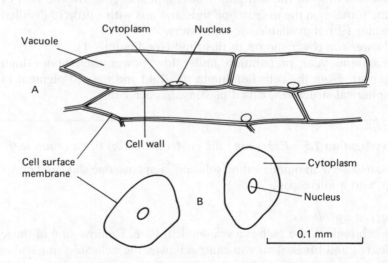

Figure 7.4 cells from: (A) a plant (epidermis of an onion bulb); and (B) an animal (epithelial cells from the skin of a mammal). (From Barrass, R. Modern Biology Made Simple)

7.3   The cell theory

You may conclude from your own observations that organisms as different as a flowering plant and a mammal are composed of cells. As you know from Investigations 7.4–7.6, animal cells differ from those of flowering plants. Animal cells do not have a cell wall; they do not have a large, central, fluid-filled vacuole; they do not contain plastids (for example, chloroplasts, see Figure 15.4). However, note particularly that the cells of all protists, plants and animals are very much alike. They all have a nucleus and cytoplasm and are bounded by a surface membrane. They also have other structures that can be seen with a microscope if suitable stains are used. Additional evidence for basic similarities between protists (for example *Amoeba* and *Chlamydomonas*), fungi (for example yeast), plants and animals, has come from studies with electron microscopes.

The word cell was used first by Robert Hooke, a British microscopist in the seventeenth century, for the empty spaces he observed in thin sections of cork. We now use the name cell for a unit of life.

Note, once again, how scientific studies began with observations that were recorded and published. In this way, scientists in different countries were able to add their observations to those made by Leeuwenhoek in Holland and Hooke in Britain. The cell came to be recognised as a basic unit of living matter. This conclusion was stated concisely in 1839 by two German scientists, Schleiden and Schwann, as the cell theory.

The cell theory is that living organisms are either single cells or are composed of many cells. That is to say, either they are unicellular

Table 7.1   *How many kingdoms?*

Kingdom	Examples	Cells
1 Animals	See Figure 3.3	
2 Protists	*Amoeba* *Chlamydomonas* seaweeds ?	Eukaryotes: unicellular or multicellular, each cell with a nucleus
3 Plants	See Figure 3.2	
4 Fungi	yeasts, moulds, mushrooms	
5 Bacteria	See Figure 7.3	Prokaryotes: unicellular; each cell with a nuclear area.
? Viruses	Influenza virus	Not cells, each with nucleic acid core.

Note. Some biologists group the seaweeds and protists in a kingdom called the Protoctista. Other biologists include the seaweeds in the plant kingdom.

(as are *Amoeba*, *Chlamydomonas* and yeast) or they are multicellular (like a flowering plant, and like yourself).

A *cell* may be defined as *a nucleus and cytoplasm bounded by a surface membrane*.

7.4   Evidence, hypotheses, theories and laws

Many of the things considered in the first seven chapters of this book have

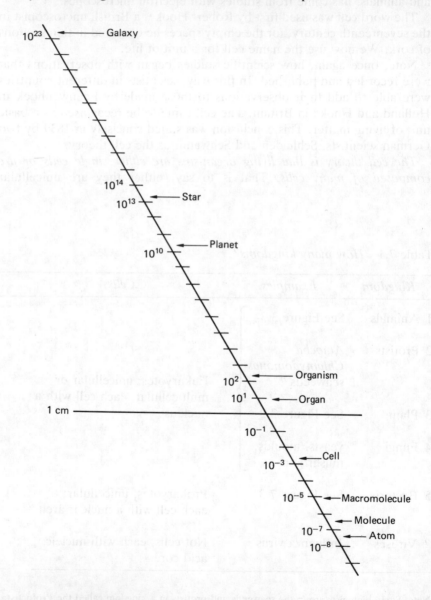

Figure 7.5 the size of things. (Based on Bonner The Scale of Nature, World's Work)

been observed and measured by scientists, or estimates have been made of their size (see Figure 7.5).

Observation is the basis of science. Ptolemy, for example, made observations and found that most of them fitted the suggested explanation that the sun, moon and planets orbited Earth (see page 17). Such a suggested explanation that seems to account for the observations is what scientists call a *hypothesis*. Ptolemy knew his simple explanation was unsatisfactory, so he put forward the additional hypothesis that each planet moved not in a circle but in a series of circles.

Copernicus suggested a simpler hypothesis, that Earth rotated on its own axis and orbited the sun. His hypothesis continues to provide a satisfactory explanation for all later observations on this subject. Such hypotheses, which *account for all known observations on a subject and which have gained general acceptance among scientists*, are called *theories*. More than this, if a theory is thought to express a basic truth it may be stated clearly and concisely as a *law of science*.

In this book you have read, for example, about Ptolemy's hypothesis, Schleiden and Schwann's cell theory, and Newton's laws of motion. Hypotheses, theories and laws are useful as summaries of what scientists think. They also help us predict what is likely to happen in future observations — if the hypothesis, theory or law is correct. But scientists must always be prepared to consider new evidence and new ideas. The laws of science must not be allowed to become dogma (see page 18) because this would hinder, not help, the progress of science.

8 The composition of matter

All materials are composed of either pure chemical substances or mixtures of chemical substances.

8.1 Chemicals

Each chemical in a science laboratory is clearly labelled with its chemical name. Gases are kept in metal cylinders under high pressure, so that a lot of gas can be stored in a small space. Liquids are usually kept in narrow-necked bottles, so they do not spill easily, even when the stopper is removed. Solids are usually kept in wide-mouthed jars with screw-on lids.

All pure chemicals must be handled carefully. If you use a pipette to draw up liquid from a bottle, do not insert this pipette into any other chemical until the pipette has been washed and dried. If you use a spatula to remove a powder from one jar, do not insert this spatula into any other chemical until the spatula has been washed and dried. By following these simple rules, you can avoid contaminating one pure chemical with traces of another. That is to say, you can avoid making what is supposed to be a pure chemical into an impure chemical.

Flammable   Explosive   Toxic

Harmful   Oxidising agent   Corrosive

Figure 8.1 chemical hazard warning labels. Each label should include a symbol and a word to indicate the nature of the hazard

Warning. All chemicals must be handled carefully. Many are toxic (poisonous) if taken into the mouth or swallowed, even in very small quantities. Some chemicals are corrosive or irritate the skin. The gases given off by many chemicals could cause permanent damage to your breathing passages, lungs and other organs, if inhaled even in very small quantities.

Every gas cylinder, bottle or jar in a science laboratory should be labelled not only with the name of the substance it contains but also to draw attention to any particular hazard (see Figure 8.1). Every chemical used in the home, garden or workplace should also be properly labelled. Never put anything into a container without labelling it properly and always remove the old label. Also, make sure all chemicals are kept out of reach of children.

When chemicals are being used, eye protection should be worn by everyone present. For many procedures a fume cupboard is essential.

In schools and colleges, chemicals should be available for use only in properly equipped science laboratories and only when supervised by a suitably qualified teacher.

8.2 Elements and compounds

Some chemical substances are called elements because they cannot be split into simpler substances by any chemical means.

An element is a substance that contains only one kind of atom. The atoms of most elements do not occur alone. They are combined in molecules (see page 141) or in larger structures (see page 143). Some molecules contain only one kind of atom, but the atoms of most elements on Earth are combined with the atoms of other elements in chemical compounds. *A chemical compound is a substance that contains atoms of two or more elements chemically united in definite proportions.*

Chemical substances present on Earth as pure elements include nitrogen and oxygen. These two gases are the main constituents of Earth's atmosphere. Water, the main constituent of lakes, rivers and seas, is a compound of hydrogen and oxygen.

Some elements do occur in a pure state in Earth's crust. Diamonds are pure carbon and gold occurs naturally as pure gold. Both are rare and have long been valued for use in jewels and other ornaments.

Little was known about elements or compounds until chemists learned how to separate the different chemical substances present in mixtures and how to separate the different kinds of atom present in chemical compounds. Only when they had sufficient quantities of a pure chemical substance could they begin to investigate its physical and chemical properties.

Obtaining pure chemicals from mixtures

A mixture consists of two or more chemical substances (elements or compounds) that are not chemically combined. It is usually possible to separate the substances by taking advantage of the fact that every chemical substance has unique physical properties. Four methods for preparing pure chemicals are chromatography, filtration, simple distillation, and fractional distillation.

Chromatography. Coloured chemicals in a solution can be separated by chromatography (Greek: *chroma* = colour). In chromatography a small spot of the mixture is placed on an absorbent material, such as filter paper. An appropriate solvent is allowed to soak through the material. As it does so, different molecules in the mixture are carried along at different rates — and in this way they separate from one another.

Investigation 8.1 *Separating the dyes present in water-soluble ink*

You will need a strip of absorbent paper (filter paper, or the edge of a newspaper will do), water–soluble ink or a felt-tipped pen, distilled water in a jar or beaker, and a pencil.

Proceed as follows
1 Put a spot of ink about 2 cm from one end of the paper.
2 Dip the paper about 1 cm into the water as in Figure 8.2.
3 Observe the movement of the water rising through the paper, separating the different dyes in the ink.
4 When the colours are well separated remove the paper from the water and dry it.

Filtration. A *filter* is used to separate either an insoluble solid from a liquid or a soluble solid from an insoluble solid. The mixture is poured on to a porous material (called a filter). The small holes in the filter allow the liquid to pass through, while the solid particles are retained.

Investigation 8.2 *Separating salt from sand by filtration*

You will need a mixture of salt and sand, distilled water, a filter funnel, filter paper and two jars or beakers.

Proceed as follows
1 Fold the disc of filter paper twice, to form a cone.

2 Fit this in the funnel, supported in a jar (as in Figure 8.3).
3 Stir the mixture of salt and sand in distilled water in the other jar.
4 Pour this mixture into the cone of filter paper. The liquid collected is called the *filtrate* and the sand remaining in the filter paper is called the *residue*.
5 Allow water to evaporate from the filtrate, leaving crystals of salt.

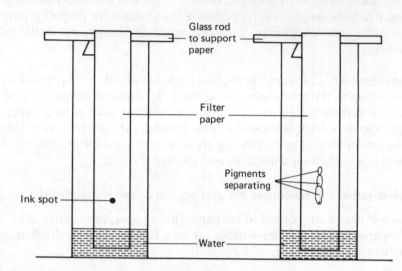

Figure 8.2 paper chromatography

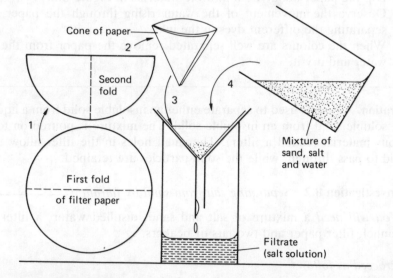

Figure 8.3 filtration

In Investigation 8.2, water (a *solvent*) is used to *dissolve* salt (the *solute*), forming a *solution* of salt in water. We say that salt is *soluble* in water:

solute + solvent = solution

The sand does not dissolve: it is *insoluble* in water.

Distillation. The solvent can be regained from a solution by distillation. First the solution is boiled so the solvent evaporates. Then the vapour is cooled so it condenses and can be collected as a liquid. The solute remains in the boiler.

Investigation 8.3 *Obtaining pure water from sea water*

Common salt (sodium chloride) and smaller amounts of other salts are in solution in sea water.

You will need sea water, the apparatus illustrated in Figure 8.4 and eye protection.

Proceed as follows
1 With the apparatus, boil the sea water gently to convert liquid water to water vapour. In the condenser, the water vapour passes through a tube surrounded by a wider tube full of cooling water. As it cools, the water vapour condenses and water (called the distillate) is collected in the beaker.
2 You will find that this water, called distilled water, is not salty. The salt does not vaporise: it is left in the flask.

Large-scale desalination plants, used to remove salts from sea water so that it can be used to drink, look very different from the small-scale laboratory apparatus but they work on the same principles: heating to vaporise, then cooling to condense.

Fractional distillation. Mixtures of liquids can be separated by fractional distillation. This technique depends on the fact that different chemical substances (pure elements and pure compounds) boil at different temperatures.

When a mixture of liquids is heated, the substance with the lowest boiling point boils first and its vapour can be condensed and collected as a pure liquid before the next substance boils, condenses and is collected in another container. Each pure liquid collected is called a fraction. The simplest method of fractional distillation is to use the apparatus for distillation (Figure 8.4), but with a thermometer fixed through the cork so

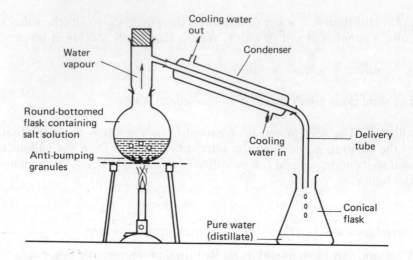

Figure 8.4 distillation apparatus used in a laboratory

the mixture in the boiler can be kept at a steady temperature while each fraction is collected.

Fractional distillation is used in industry, for example, to separate alcohol (ethanol boils at 78 °C) from water (boiling point 100 °C) and to extract different chemicals from crude oil (see page 291).

Elements

Although Democritus suggested 2400 years ago that all matter was composed of very small particles, other Greek philosophers, including Aristotle, argued that all matter was continuous (not composed of particles).

The existence of different elements, each a basic substance that could not be split into other substances but could combine with other elements to form compounds, was suggested by Robert Boyle, an English pioneer of modern chemistry, in 1661.

Two French chemists who made contributions were Lavoisier and Proust. In the 1770s Lavoisier listed 33 substances he thought were elements and Proust proved that the elements in a compound were present in definite proportions.

In 1803 another English chemist, John Dalton, pointed out that much of what was known about elements and the way they combined to form compounds could be understood if each *element* were *composed of only one kind of atom*, with unique physical and chemical properties.

If this *atomic theory* were correct, the atoms of different elements would be expected to differ in mass (see page 34). Chemists started to compare the atomic masses of different elements. Then they placed them in order of atomic mass, starting with hydrogen — the lightest element.

When bromine (a liquid) was discovered in 1826, it was found to have properties mid-way between those of chlorine (a gas) and iodine (a solid). When the atomic mass of bromine was determined, three years later, it was very close to the arithmetic mean of the atomic masses of chlorine and iodine.

Using present-day figures, based on an atomic mass of 16 for oxygen, the relative atomic mass of chlorine is 35.453 and that of iodine is 126.904 (Sum = 162.357). This, divided by 2, gives an arithmetic mean of 81.179 which is close to the atomic mass of bromine (79.909).

Other sets of three elements were known in 1826 (for example, the metals lithium, sodium and potassium) in which the atomic mass of the middle element was half-way between the atomic masses of the other two. But no one at the time appreciated the importance of these sets of three.

In 1866 a young English chemist, John Newlands, arranged elements in eights, like the octaves of music, and noted that this brought elements with similar properties together in groups of three. Other chemists laughed at this musical arrangement. However, a similar arrangement, called a *periodic classification of elements* because of the *regular (periodic) arrangement of elements with similar properties*, was proposed by Mendeleev, a Russian chemist, in 1869. He arranged the elements with similar properties together even if this meant he had to leave gaps in his table. Indeed, in leaving these gaps Mendeleev went further. He (1) suggested that elements would be discovered which fitted into these gaps and (2) predicted what the properties of these elements would be.

Table 8.1 *Part of the periodic classification of elements*

Eight groups of elements							
1	2	3	4	5	6	7	0
Hydrogen $_1$H							Helium $_2$He
Lithium $_3$Li	Beryllium $_4$Be	Boron $_5$B	Carbon $_6$C	Nitrogen $_7$N	Oxygen $_8$O	Fluorine $_9$F	Neon $_{10}$Ne
Sodium $_{11}$Na	Magnesium $_{12}$Mg	Aluminium $_{13}$Al	Silicon $_{14}$Si	Phosphorus $_{15}$P	Sulphur $_{16}$S	Chlorine $_{17}$Cl	Argon $_{18}$Ar
Potassium $_{19}$K	Calcium $_{20}$Ca					Bromine Br	Krypton Kr
						Iodine I	Xenon Xe
METALS			**NON-METALS**				**INERT GASES**

Scientists now know of 109 elements. They have been arranged in order of increasing atomic mass, and given an atomic number. Hydrogen (atomic number 1) is the lightest and plutonium (atomic number 94) is the heaviest naturally occurring element. The remaining 15 elements have been manufactured.

Each element has been given its own internationally accepted symbol. For example, H is the symbol for one atom of hydrogen. Each symbol is an abbreviation of the element's Latin name, and so is not necessarily the initial letter of the element's English name.

The first twenty elements of the periodic classification, with their atomic numbers and chemical symbols, are arranged in Table 8.1 in order of increasing atomic mass but in horizontal rows so that groups of elements with similar properties are in vertical columns.

This account of the discovery, orderly arrangement and classification of chemical elements illustrates the importance of observation, orderly arrangement, classification and prediction in science. It also illustrates the importance of communication in science, so scientists of all nations can cooperate in contributing to the advancement of scientific knowledge and understanding.

Chemical compounds

The elements that combine to form a compound have different properties from the compound. For example, water (a liquid) is composed of hydrogen and oxygen (two gases). Edible common salt (chemical name: sodium chloride) is a compound of the element sodium (a dangerous metal that can react explosively) and the element chlorine (a poisonous gas).

The elements that make up a compound are indicated by their letter symbols in the *chemical formula* of the compound. For example, the chemical formula for sodium chloride (NaCl) indicates that in sodium chloride there is one atom of sodium (Na) for every atom of chlorine (Cl). The chemical formula for sulphuric acid (H_2SO_4) indicates that in a molecule of sulphuric acid two atoms of hydrogen (H), one atom of sulphur (S) and four atoms of oxygen (O) are chemically united in these definite proportions.

Chemical change. *A change in the chemical composition of substances*, as when elements combine to form a chemical compound, is called a *chemical change*. This results in a change in the number or in the arrangement of atoms and new materials are formed. Such a change cannot be reversed easily. Contrast this with a physical change (see page 36) in which there is no change in the composition of a substance (no new materials are formed) and which may be reversed easily (see page 39).

A chemical change is the result of a reaction between different molecules. This is called a *chemical reaction*. It can be represented in words or in symbols as a *chemical equation*. For example:

Two molecules of hydrogen	and	One molecule of oxygen	combine	Two molecules of water
$2H_2$ (g)	+	O_2 (g)	$\longrightarrow$	$2H_2O$ (l)

The number in front of the chemical formula for a molecule is the minimum number of molecules combining to give the correct proportions. If there is no number this means there is just one molecule. The letters in brackets in such equations are: g = gas, l = liquid and s = solid.

Note that all the atoms on the left side of a chemical equation (the reactants) must also be present on the right side (the products). That is to say, the equation must balance. The reason for this is that *in a chemical reaction, matter can neither be created nor destroyed*. This is the *Law of Conservation of Matter*, also known as the *Law of Conservation of Mass* (see page 335).

Each chemical compound has unique physical and chemical properties, by which it can be distinguished from all other chemical substances (elements and compounds). Some compounds can be produced in more than one way but scientists have found that however it is made: *a particular kind of chemical compound always contains the same elements combined in the same proportions*. This is another law of science: the *Law of Constant Composition*.

⑨ Cells and tissues

In the growth of a flowering plant from a seed, the seedling grows at the shoot tip, producing new leaves at intervals on the stem and an axillary bud in the axil of each leaf (see Figure 5.5). It also grows at the root tip.

Investigation 9.1 *The growth of a root tip*

You will need a broad bean seed, a glass tube or jar, cotton wool, blotting paper, water, a mm rule, a mapping pen and red ink.

Proceed as follows
1 Allow a broad bean seed to germinate, as in Investigation 5.2.
2 When the root is about 2 cm long, mark it with dots of red ink at 1 mm intervals from the tip. Then arrange it in a tube (as in Figure 9.1).
3 Keep the tube in the dark overnight. Then measure the distance between successive ink marks. Record the measurements in your notebook.
4 Prepare a graph (similar to the graph in Figure 9.1) based on your own data. Record your conclusions about growth at the tip of this root.

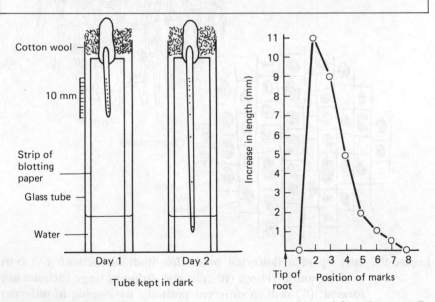

Figure 9.1 growth of a young root of a broad bean *Vicia faba.* **(Barrass, R.** Modern Biology Made Simple)

9.1 Cells near a root tip

If you examine a thin slice cut along the length of a root tip (see Figure 9.2) you will see that cells in different positions differ in size and shape and that the smallest cells are near the tip. Such thin sections can be cut with a microtome, which works like a bread slicer.

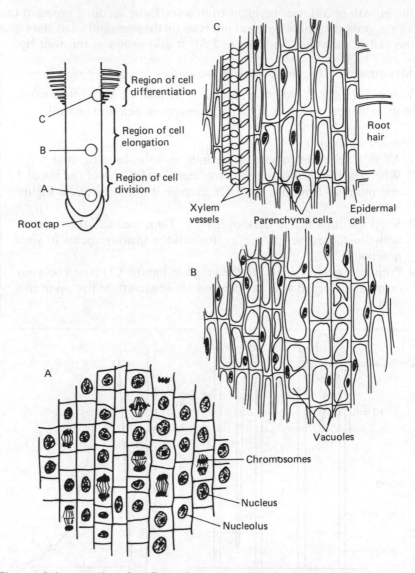

Figure 9.2 root tip of a flowering plant (longitudinal section): (A) cells growing and dividing; (B) cells elongating as large vacuoles are formed; (C) cells in different positions developing in different ways (differentiating). Diagram not to scale. (From Barrass, R. Modern Biology Made Simple**)**

Investigation 9.2 *Preparing a squash of a root tip of an onion*

You will need a glass jar, an onion that will rest on the top of the jar, water, a razor blade, two watch-glasses, a microscope slide and cover slip, concentrated hydrochloric acid, absolute alcohol, 45 per cent glacial acetic acid, acetic orcein stain, blotting paper and a microscope.

Warning Acid and bases are corrosive. Wear eye protection when handling acids and bases. They attack clothing and flesh and should be handled with great care. Any splashes should be washed away immediately with plenty of water.

When diluting an acid with water, always add the acid slowly to the water, stirring all the time. *Never add water to an acid.* To do so would cause dangerous spitting.

Proceed as follows

1 Rub any loose material from the base of the onion then place the onion on a jar of water, so the base of the onion is about 1 mm above the water. The roots that grow from the base of this bulb are suitable for a root tip squash preparation.
2 Wearing eye protection, cut off a root tip, about 1 cm long, and place it in a watch glass containing equal volumes of concentrated hydrochloric acid and absolute alcohol. Leave for 5–10 min.
3 Transfer the root tip to another watch-glass containing 45 per cent glacial acetic acid. Leave for 5 min.
4 Place the root tip on a microscope slide and add a drop of acetic orcein.
5 Cut the tip in half, along its length, and similarly into quarters, then keep on cutting lengthwise until you have many long strips. Do not let them dry. If necessary add another drop of stain.
6 Place a cover slip on your preparation.
7 Fold blotting paper (or filter paper) into a thick pad. Place this over the cover slip. Then with one thumb on this pad, use your other thumb to press gently on the preparation to squash it. This should separate the cells without disrupting them.

Acetic orcein stains chromosomes (see Figure 9.2A) in cells that are dividing or are just about to divide but it does not stain other parts of these cells. In your preparation you will see many cells, outlined by their cell walls. Note the following things about chromosomes (Greek: *chroma* = colour; *soma* = body).

Chromosomes

(1) You cannot see chromosomes in all the cells. (2) All the cells in which you can see chromosomes are together in the region just above the root tip

(see Figure 9.2A). (3) All the cells in which you can see chromosomes have the same number of chromosomes. (4) This number of chromosomes is an even number.

If you repeat this squash preparation with the root tips of any other plant you will make the same observations. The explanation is as follows: (1) It is possible to stain chromosomes only in cells that are either about to divide or are dividing. (2) All the dividing cells in a young root are together in the region just above the root tip. (3) All the body cells of any plant or animal have the same number of chromosomes (for example: fruit fly = 8, mouse = 40, man = 46). (4) If you find one chromosome that has a particular size and shape in a cell, you can find another that is the same size and the same shape in the same cell. The chromosomes differ in size and shape but there are always two of each kind.

9.2   Cell division

The cell division which results in more body cells being produced, called mitosis, is similar in all plants and animals (see Figure 9.4) except that the cells of plants have cell walls and do not have centrioles.

At the start of mitosis the chromosomes can be stained. They are long thin structures in the nucleus. As they become shorter, pairs of chromosomes can be recognised. The two centrioles migrate, outside the nucleus, to the poles of the nucleus. The nuclear membrane and the nucleolus disappear and a spindle of fibres forms between the centrioles.

The chromosomes are at the equator of the cell and *each chromosome comprises two chromatids*.

In the separation stage *the chromatids of each chromosome move apart* — one to each pole of the spindle (see Figure 9.3). During this movement the

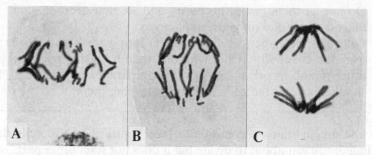

Figure 9.3 three cells in the root tip of a flowering plant *Haworthia cymbiformis* **(magnification × 975) in the separation stage of mitosis: (A) chromosomes (8 long and 6 short in this species) at equator of cell but with chromatids moving apart; (B) chromatids further apart; (C) chromatids at poles of spindle. (Preparation and photographs by Dr P. Brandham, Royal Botanic Gardens, Kew, England)**

chromatids appear to be attached to the spindle by their centromeres. The chromatids become longer until, once again, they cannot be stained. A nuclear membrane forms around each set of chromatids and a nucleolus in each nucleus. Then the cytoplasm divides.

After a while one or both of the cells may divide again, but before this happens more cytoplasm is produced, a duplicate of each chromatid is formed and the centriole is duplicated. As a result, at the start of the next mitosis each cell is a copy of the cell from which it was formed.

In plants, after nuclear division, each new cell secretes a cell wall which divides the cytoplasm roughly equally between the two new cells.

The result of repeated cell division, in both plants and animals, is that the number of cells in the body increases and all the cells of each multicellular organism have an identical set of chromosomes.

Chromosome structure

Each chromosome includes long molecules of protein, called nucleoprotein. Each of these protein molecules acts as a carrier of deoxyribonucleic acid (DNA) molecules. Each nucleic acid molecule is a chain of nucleotides (Figure 9.5A) with deoxyribose sugars, linked by phosphoric acid molecules, and with either a purine molecule (adenine = A; or guanine = G) or a pyrimidine molecule (thymine = T; or cytosine = C) attached to each sugar.

In any organism, there are always equal numbers of A and T and always equal numbers of G and C. This is what we should expect if these molecules always occur in pairs A-T and G-C (see Figure 9.5B). This line of reasoning led Watson and Crick to suggest in 1953 that the long molecule of DNA comprised two chains wound around one another and forming a double helix (see Figure 9.5C).

Chromosome duplication

The duplication of the DNA molecule is thought to take place following the unwinding of the spiral, which breaks the linkages A-T and G-C (see Figure 9.5 D and E). Where an A is exposed, a T is added from the nuclear fluids, and where a T is exposed an A is added. In a similar way, an exposed G can link only with a C, and an exposed C only with a G. In this way, the unwinding of the double helix in the nucleus which contains the necessary raw materials, is accompanied by the construction of two new chains (Figure 9.5F). These two chains are identical. They are also replicas of the unwinding double helix that formed the templates for their construction.

This copying of the DNA molecules is the basis for the duplication of the nucleoproteins, and therefore for the duplication of chromatids (see above) between the end of one cell division and the start of the next. This results in all the body cells of any one multicellular organism having the

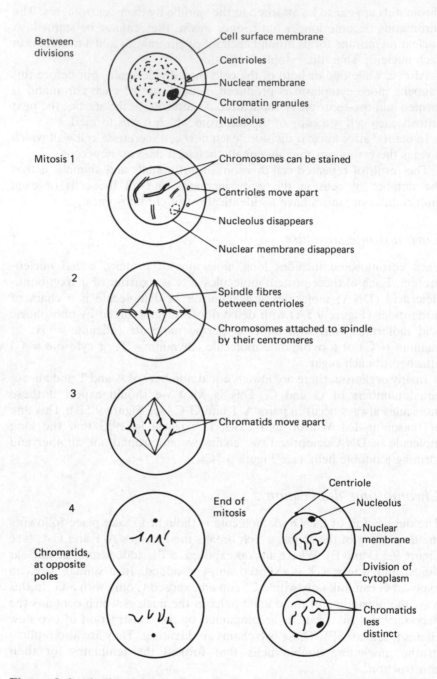

Between divisions — Cell surface membrane
— Centrioles
— Nuclear membrane
— Chromatin granules
— Nucleolus

Mitosis 1 — Chromosomes can be stained
— Centrioles move apart
— Nucleolus disappears
— Nuclear membrane disappears

2 — Spindle fibres between centrioles
— Chromosomes attached to spindle by their centromeres

3 — Chromatids move apart

4 — End of mitosis
Chromatids, at opposite poles
— Centriole
— Nucleolus
— Nuclear membrane
— Division of cytoplasm
— Chromatids less distinct

Figure 9.4 mitotic division of a cell of an animal. Diagram not to scale. (From Barrass, R. Modern Biology Made Simple**)**

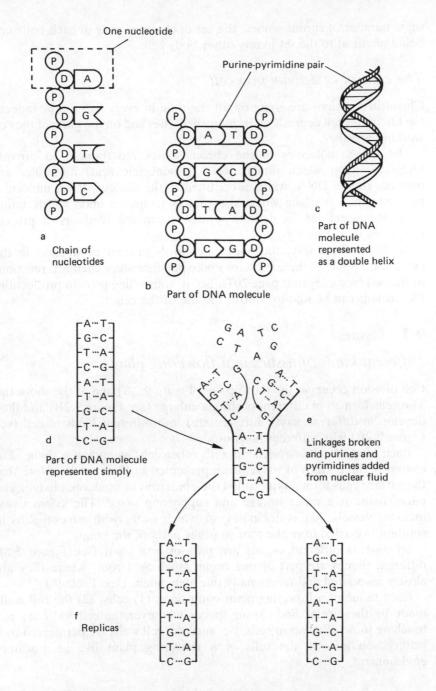

Figure 9.5 The DNA molecule: (a to c) structure; (d to f) method of duplication (or replication). Key to symbols: D = deoxyribose sugar; P = phosphate; A and G = purines; T and C = pyrimidines (From Barrass, R. Modern Biology Made Simple)

same number of chromosomes, the set of chromosomes in each body cell being identical to the set in any other body cell.

The control of activity in a cell

Chemical reactions are going on all the time in every living cell. Indeed, the life of the cell depends on its many activities and on the parts of the cell working together.

The DNA molecules of the chromosomes are thought to provide templates upon which similar RNA (ribonucleic acid) molecules are formed. Unlike DNA, which is confined to the nucleus, RNA molecules pass into the cytoplasm where they act as templates upon which amino acids are assembled, in definite sequences, in the synthesis of protein molecules (see page 196).

In this indirect way, the nucleus controls protein production in the cytoplasm. Some of these proteins make possible other chemical reactions in the cell (see enzymes, page 207). So, in controlling protein production, the nucleus can be said to control the life of the cell.

9.3   Tissues

Different kinds of tissues in a flowering plant

Cell division occurs at the root tip (see Figure 9.2A) and at the shoot tip. The cells formed in different positions enlarge (see Figure 9.2B) and then develop in different ways (differentiate) into different kinds of cell (see Figure 9.2C) with different functions.

Each kind of cell is associated with other cells as part of a tissue. For example, three kinds of tissue are represented in Figure 9.2C. Note that the surface cells form a skin, called the epidermis or epidermal tissue. The parenchyma is a water storage and supporting tissue. The xylem tissue includes vessels, like water pipes, in which water with mineral salts in solution is carried from the root to other parts of the plant.

So, just as different organs are part of one plant (see Figure 5.5), different tissues are part of one organ (such as a root) where they are closely associated and functionally interdependent (see Table 9.1).

Each tissue of a flowering plant comprises: (1) cells, (2) the cell walls made by these cells, and (3) air spaces wherever the cell walls are not touching those of adjacent cells. Because the cell walls are permeated by a watery solution, all the cells of a flowering plant live in a watery environment.

Different kinds of tissues in the body of a mammal

Your body comprises: (1) cells, (2) materials made by these cells, and (3) fluids that bathe the cells. Like the cells of a flowering plant, and those of other organisms, the cells of your body live in a watery environment. Some

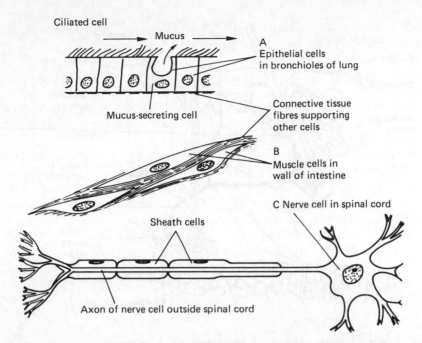

Figure 9.6 cells in different tissues of your body. Diagram not to scale. (From Barrass, R. Human Biology Made Simple)

cell types, each with different functions, are illustrated in Figure 9.6 to introduce you to three kinds of tissue: epithelium, muscle and nerve.

Table 9.1 *Interdependence of some tissues in a flowering plant*

Tissue	Some functions
Epidermis	Acts as a skin, supports and protects internal tissues, intake of water and nutrient ions through root
Parenchyma	Water storage and supporting tissue
Xylem	Transports water with mineral ions in solution from root to all other parts of plant, also gives mechanical support
Phloem	Transports organic molecules produced in leaves to all other parts of plant
Cambium	Dividing cells in shoot and root produce more xylem and phloem as the plant gets bigger.

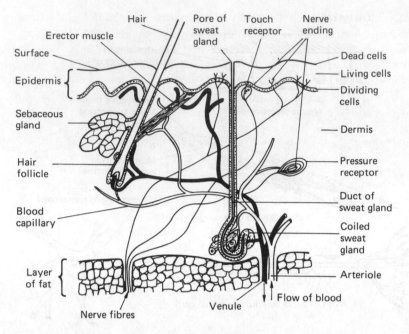

Figure 9.7 vertical section of the skin of a man or woman. (From Barrass, R. Human Biology Made Simple)

The tissues of your body, like those of a flowering plant, are closely associated in each organ, and functionally interdependent (see Table 9.2).

Your skin: tissues working together. Your skin is mainly composed of two tissues (see Figure 9.7): a surface epithelium (the epidermis) and connective tissue (the dermis). Note that the dermis also contains blood capillaries (epithelial tubes) and nerves (nerve tissue).

The epidermis is called a compound epithelium because it is more than one cell thick. The outer cells are formed as a result of the repeated growth and division of cells in the basal layer. Recently formed cells, near the basal layer, are alive. Older cells fill with a hard protein (keratin) and die, forming a tough outer layer that protects your internal tissues.

The basal layer of the epidermis extends into the hair follicles and into glands. Some cells in each hair follicle divide at the base of a hair and some new cells become keratinised — making the hair longer. In sebaceous glands some cells divide, and some new cells fill with a secretion called sebum. As these cells disintegrate the sebum lubricates the hairs and helps to keep the skin supple. It also has bactericidal properties. Cells at the base of each finger nail and toe nail divide and some new cells become keratinised. So your nails grow by the addition of new material at the base, and you keep them short by cutting.

The dermal connective tissue: (1) supports the epidermal structures (epidermis, hair follicles, sebaceous glands and sweat glands), (2) supports the many blood capillaries and nerves that run through it, and (3) with the epidermis supports and protects the internal organs and helps to maintain the shape of the body. Also, (4) some connective tissue cells contain stored fat (see Figure 9.7) and form both an insulating layer (see page 239) and a food reserve (see page 209).

Table 9.2 *Interdependence of tissues in the body of a mammal*

Tissue	Some functions
Epithelium	Covers surfaces and lines tubes, always protects, in some places materials pass through into or out of the body
Connective	Supports other tissues
Muscle	Contracts
Nerve	Conducts nerve impulses
Blood-forming	Production of white cells, erythrocytes and platelets

10 Energy, work and power

The universe consists of matter and energy. Science, therefore, is the study of both matter and energy. In contrast to matter, which occupies space and has mass, energy does not occupy space and it has no mass. *Energy* may be defined as *that which is used when work is done* or as *the capacity to do work*.

10.1 Forms of energy

Mechanical energy, heat energy, chemical energy, light energy and electrical energy are different forms of energy. Even though energy occupies no space and has no mass, you are aware of these different forms of energy.

The parts of your body that are sensitive to energy in your surroundings are called *sense organs* (see Table 10.1). Touch, hearing, taste, smell and sight are your five *senses*. The forms of energy that affect your sense organs are called *stimuli*.

Table 10.1 *Our sensitivity to different forms of energy*

Sense organ	Sense	Stimulus
Skin	Temperature	Heat energy
	Touch	
Ears	Hearing	Mechanical energy
Eyes	Sight	Light energy
Tongue	Taste	Chemical energy
Nose	Smell	

Mechanical energy

When the lid of the toy illustrated in Figure 10.1 is closed, work is done in compressing the spring. At the same time, mechanical energy is stored in the spring. This is called stored or *potential energy*. If the catch on the lid is released, the doll pops up. The potential energy that was stored in the spring has been changed to movement energy that is doing the work. Another name for this movement energy is *kinetic energy*.

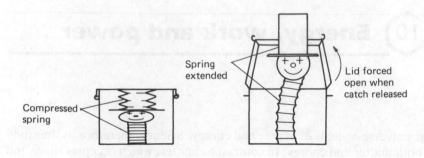

Figure 10.1 jack-in-a-box

So you see that mechanical energy may be either stored in an object as a result of its position (the position of a force, see page 57) or used to do work when an object moves (when the force moves).

There are many everyday occurrences in which potential energy is changed to kinetic energy. For example, an object gains potential energy when it is raised and this stored energy is changed to kinetic energy when the object is allowed to fall. Such energy changes occur repeatedly as a pendulum swings to and fro. As a pendulum bob is raised it gains potential energy, so energy is stored in the bob.

When the pendulum is released (for example at point A in Figure 10.2) it gathers speed as most of its potential energy is changed to kinetic energy. At the lowest point in its swing (B in Figure 10.2), the pendulum has no potential energy but it continues to swing as far as point C (in Figure 10.2), slowing down as kinetic energy is changed to potential energy. At point C it has no kinetic energy but again has potential energy because of its raised position.

Note that point C in Figure 10.2 is not as high as point A, because some energy is used to push the pendulum against the resistance of the air and some energy is wasted in rubbing (friction) at the point of suspension. That is to say, some of the potential energy the pendulum had at point A is not changed to kinetic energy, and some of the kinetic energy it had at point B is not changed to potential energy. If they were, point C would be as high as point A and the pendulum might be expected to go on swinging forever. In practice the length of the swing gradually decreases until the pendulum eventually stops moving.

Whenever there is friction, some movement energy (mechanical energy) is changed to heat energy. You can feel this when you rub your hands together. In any energy–changing arrangement, such as a pendulum swinging, some energy is wasted in the sense that it is not available to do useful work. But remember that energy is not destroyed; as the amount of mechanical energy decreases, the amount of heat energy increases. Scientists say that energy is conserved. This is another law of science: the *law of*

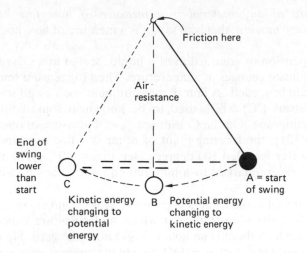

Figure 10.2 energy changes as a pendulum swings

conservation of energy: Energy can be converted from one form to another, but energy can neither be created nor destroyed.

Compare this law with the law of conservation of mass (or matter) (see page 89). As we shall see, with increasing knowledge of matter and energy these two laws have been combined in the law of conservation of mass and energy.

Heat energy

People have long used fuels as sources of heat, even when they had no idea what heat was. Then in the 1840s James Joule, an English scientist, measured accurately the amount of heat produced when work was done. He found that a given quantity of mechanical energy always produced an identical amount of heat. Joule concluded that heat was a form of energy.

The kinetic theory (see page 40) states that all matter is composed of molecules in motion. *Heat energy is the energy of moving molecules.* As a material is heated, the input of energy causes its molecules to move faster and come to be further apart. Heating, therefore: (1) results in expansion, and (2) may result in a change of state (see page 37). Heat may also: (3) cause a rise in temperature, (4) have a chemical effect (see page 108), and (5) have an electrical effect which is not considered in this book. These are the five effects of heat.

When heat is absorbed by any material its molecules move faster: we say it gets hotter (unless there is a change of state, see page 107). *The temperature of any material is a measure of how fast its constituent molecules are moving:* that is to say, it is a measure of how hot the material is.

The expansion or contraction of a liquid, sealed in a glass tube, can be used to indicate changes in temperature. Then to measure temperature an agreed scale is needed. A unit of measurement used by all scientists is the degree Celsius (°C) which used to be known in some countries as the degree Centigrade. On the Celsius scale, at a pressure of one atmosphere (see page 121), the freezing point of water is called zero and the boiling point of water is called 100 degrees (see Figure 10.3). The thermometer scale is therefore marked by a hundred equal divisions between 0 °C and 100 °C.

As a material cools its molecules move slower and slower. In the 1860s Kelvin called the temperature at which molecules are expected to stop moving (at which there is no heat energy) absolute zero. He chose this as the zero point (zero kelvin = 0 K) on a new temperature scale (the Kelvin scale, see Figure 10.3). Absolute zero has been calculated as approximately −273 °C, but such a low temperature has never been achieved.

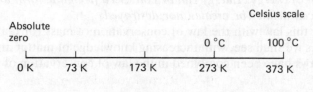

Figure 10.3 temperature scales. Note: the Fahrenheit scale (32 °F = 0 °C; 212 °F = 100 °C) is no longer used by scientists

---

Investigation 10.1 *Temperature change as ice melts*

You will need crushed ice, a thermometer, and a jar or beaker.

Proceed as follows
1 Pack ice in the jar around the thermometer.
2 In a table in your notebook, record the temperature at 1 min intervals until all the ice has melted and the temperature is rising steadily.
3 Prepare a time/temperature graph, with time on the horizontal axis (min) and temperature on the vertical axis (°C)

As ice absorbs heat from the room, its temperature rises steadily until at 0 °C it stops rising. Then, when all the ice has melted the temperature starts to rise again. Heat energy is being absorbed all the time but for a while the temperature does not rise. Heat energy seems to disappear and is called hidden or *latent* heat.

When a solid changes into a liquid or a liquid into a gas, latent heat energy is needed to move the molecules further apart. When the change of state is in the opposite direction (gas to liquid; liquid to solid) the molecules move closer and latent heat is released.

Investigation 10.2 *The difference between heat and temperature*

You will need a stop clock, two bunsen burners or low voltage immersion heaters, two tripods, two thermometers (-10 to $110\,°C$), two $250\ cm^3$ beakers, a measuring jar or cylinder and eye protection.

Proceed as follows
1 Put $100\ cm^3$ water in each beaker. Place each beaker on a gauze, on a tripod.
2 Adjust the bunsens, so they have identical small flames.
3 Record the temperature of the water in each beaker.
4 Place a bunsen under each tripod. Repeat the temperature measurements at 1 min intervals until the water temperature in one beaker is $50\,°C$. Record your readings in a table.
5 In a graph, plot temperature (y axis) against time (x axis).
6 Now heat different volumes of water ($100\ cm^3$ in one beaker and $200\ cm^3$ in the other), using the above procedure, until the water in the $100\ cm^3$ beaker is $50\,°C$. Record your data in a table.

With $100\ cm^3$ water in each beaker, assuming they absorb identical amounts of heat energy, you would expect the rise in temperature in the two beakers to be identical. You would expect the rise in temperature of $200\ cm^3$ water to be about half that of $100\ cm^3$ water.

What did you find? You may conclude that it is not possible to measure the amount of heat used simply by measuring the temperature rise. For all energy calculations you must know the mass of the material being heated.

The increase in temperature produced by a certain quantity of heat energy also depends on the type of material being heated. For example, to raise the temperature of water, you would need about ten times as much heat as you would need to raise the temperature of the same mass of copper by the same amount. *Each material has a specific capacity for heat.*

To measure the heat energy gained or lost by any material it is necessary to know the mass of material, its specific heat capacity and the temperature change.

$$\begin{matrix} \text{Heat energy gained} \\ \text{or lost} \end{matrix} = \text{mass} \times \begin{matrix} \text{specific heat} \\ \text{capacity} \end{matrix} \times \begin{matrix} \text{temperature} \\ \text{change} \end{matrix}$$

Heat energy and work. Energy is that which is used when work is done (see page 103). In everyday language we speak of manual work and mental work, but in science for work to be done there must be a force (see Chapter 6) and movement. *Work is the use of energy to move a force.*

To calculate the energy used in any activity, that is to say the work done, you must know the size of the force and the distance moved.

$$\text{Energy used or work done} = \text{force} \times \text{distance moved}$$

The SI unit for force is the newton (see page 61) and distance is measured in metres. The SI unit for work done (energy used) is the joule (symbol J). *One joule is the work done when a force of one newton moves one metre.*

Example. How much work is done when a bag of sand weighing 10 N is lifted 15 m?

$$\text{Work done (joules)} = \text{force (newtons)} \times \text{distance (metres)}$$
$$10 \times 15 = 150 \text{ J}$$

We can now define the *specific heat capacity of any substance* as the *amount of energy needed to raise the temperature of 1 kg of the substance by 1 °C.* For example, to raise the temperature of 1 kg water by 1 °C, 4200 J of heat energy are needed. The specific heat capacity of water is 4200 joules per kilogram per degree Celsius. Note that when the temperature of 1 kg water falls by 1 °C the amount of heat energy lost is also 4200 J.

Chemical energy

Magnesium ignites spontaneously in air, giving out a brilliant light and great heat. This is a chemical reaction between the elements magnesium and oxygen (the reactants), in which magnesium oxide (the product) is formed.

$$2Mg \text{ (s)} + O_2 \text{ (g)} \longrightarrow 2MgO \text{ (s)} + \text{Light} + \text{Heat}$$

| Magnesium | Oxygen | Magnesium oxide | energy | energy |

In contrast, when iron filings and powdered sulphur are mixed they do not combine unless heated. Heat energy is needed to start the reaction:

$$\text{Iron} + \text{Sulphur} + \text{Heat} \longrightarrow \text{Iron (II) sulphide}$$

| filings | powder | energy | glows red as heat released |

Heat is needed to start many chemical reactions: this is the chemical effect of heat.

Most combustible materials (materials that will burn) have to be ignited. An input of heat energy is needed to start the chemical reaction but then much more heat energy is given out. For example:

Carbon + Oxygen $\longrightarrow$ Carbon dioxide + Heat and light
(coke)   from air   added to air   energy

Once a fuel has started to burn its potential energy (chemical energy) is converted to heat energy and light energy. Note that oxygen is needed for *combustion* (burning), and fire extinguishers work by depriving a fire of air.

As we shall see (page 170), chemical energy stored in your food and in some drinks is the only source of energy available to do work in your body. The energy content of a food can be determined by burning a known mass of the food in air, using it as a fuel to heat a known mass of water (4.2 J of heat energy are needed to raise the temperature of 1 g of water by 1 °C).

All *chemical reactions* are accompanied by some energy change. Those *in which energy is given out* are called *exothermic reactions*, and those *in which energy is absorbed* are called *endothermic reactions*.

Light energy

Your eyes are sensitive to light energy. You see some objects, including the sun, because they are luminous. And you see others, including the moon, when they are illuminated and light bounces off them (is reflected).

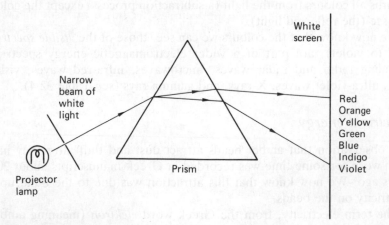

Figure 10.4 a spectrum from a beam of white light

┌─ **Investigation 10.3**   *The colours of a rainbow* ─────────────────┐

You will need a slide projector, a card cut the size of a projector slide with a small hole punched in its centre, a triangular glass prism, a white screen and a darkened room.

Proceed as follows
Put the card in the projector so a narrow beam of light falls on the screen. Switch off the room light and hold the prism as in Figure 10.4.

└──┘

Sunlight is colourless but we call it white light. Isaac Newton was the first to show, in 1665, that white light was in fact a mixture of red, orange, yellow, green, blue, indigo and violet lights.

Newton also demonstrated that the colours came from the light, not from the prism. He did this by arranging for just one of the colours to pass through a second prism; a spectrum was produced only from white light.

Newton showed that white light could be produced by passing all the coloured lights of the spectrum through a second prism held upside down. White light can also be obtained by mixing the primary colours (RED, GREEN and BLUE), or the secondary colours (turquoise, magenta and yellow), or two complementary colours (RED and turquoise, GREEN and magenta, or BLUE and yellow), (see Figure 10.5).

By mixing the primary colours in different proportions it is possible to obtain all colours. Colour televisions work by mixing lights and colour photography involves the use of a light–sensitive film that is in three layers, each sensitive to only one primary colour.

Do not confuse the mixing of lights (an addition process) with the mixing of pigments. A coloured object or substance, like pigment or paint, absorbs all colours from the light (a subtraction process) except the colour you see (the reflected light).

We now know that the colours we can see, those of the *visible spectrum* (red to violet), are part of a wider electromagnetic energy spectrum, including radio and radar waves, microwaves, infra-red waves, visible light, ultra-violet waves, X-rays, and gamma rays (see Table 32.4).

Electrical energy

The observation that amber beads attract dust and fluff after they have been worn for some time was recorded in Greek manuscripts about 2000 years ago. We now know that this attraction was due to the build up of electricity on the beads.

The term electricity, from the Greek word *elektron* (meaning amber) was first used by an English scientist, William Gilbert, in about 1600. By rubbing against clothes the amber beads became electrically charged.

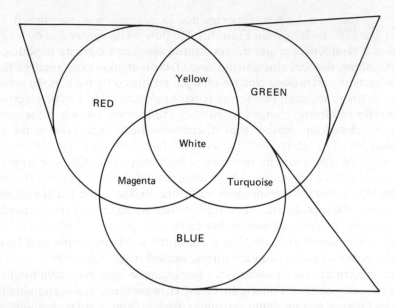

Figure 10.5 overlapping spotlights: one red, one green and one blue

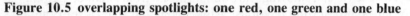

Investigation 10.4 *Charging materials by friction*

You will need pieces of tissue paper (less than 1 cm²), objects made of different materials (for example, a plastic pen, a glass rod, a smooth piece of wood, a polystyrene tile, a plastic bag), a wool cloth. Keep all the materials dry. **Warning:** use only smooth objects.

Proceed as follows
1 Hold the pen over pieces of tissue paper. Record what happens.
2 Rub the plastic pen vigorously with the cloth. Repeat step 1.
3 Repeat steps 1 and 2 with each of the other objects.

You will find that some objects, after being rubbed with the cloth, attract the pieces of paper. Other objects, especially those made of synthetic materials, attract and then repel the paper alternately: the paper seems to jump up and down.

Some materials are easier to charge than others, but all objects can be electrically charged. In a thundercloud, the water droplets become electrically charged; flashes of lightning are electrical charges suddenly leaking to earth.

One of the best known experiments in science was performed in America in 1752 by Benjamin Franklin. He flew a kite suspended by a silk thread in a thunderstorm and demonstrated that the electricity conducted to earth along the wet thread (moving electric charges) behaved in the same way as the stationary electric charges produced by friction. In other words, although we need two terms (current electricity and static electricity) to refer to electric charges when they are moving and when they are stationary, Franklin proved that there was really only one kind of electricity.

Soon afterwards, Franklin invented a lightning conductor — a strip of metal attached to a tall building. Lightning conductors help prevent flashes of lightning — and if lightning does strike the building, the metal reduces or prevents damage to the building by conducting electricity to earth. Lightning causes great economic losses as a result of striking trees and buildings not protected by lightning conductors. Many people and farm animals, out-of-doors in thunderstorms, are killed by lightning.

Static electricity can be a nuisance. For example, you may have noticed how dust accumulates on plastic surfaces. However, use has been made of static electricity in removing carbon particles from smoke, helping to reduce air pollution from factories. Also, about 200 years after Franklin's invention of the lightning conductor, in 1960 another American, Chester Carlson, developed the technique of xerography, in which carbon black is attracted to paper as a result of localised electrostatic attraction. This is the process we now call photocopying.

10.2 Energy transfer

When we use energy the amount of energy in the universe does not decrease. Because energy is not used up, it is better to speak of energy change from one form to another or energy transfer rather than energy use. Many energy transfers have been mentioned in this chapter. Some that occur in an internal combustion engine are summarised in Figure 10.6.

In any energy-changing arrangement some energy is wasted. For example, in a car engine about 85 per cent of the chemical energy in fuel is wasted — that is, it is not available to do useful work. Only about 15 per cent is converted to mechanical energy. A car engine works best if it is not too cold, but a cooling system is needed to dissipate all the excess heat energy.

10.3 Energy, work and power

Energy has been defined as that which is used when work is done, and work as the use of energy to move a force. In science we also need to

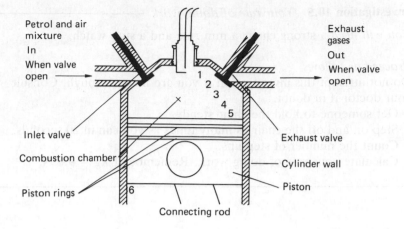

Petrol and air mixture

In

When valve open →

Exhaust gases

Out

When valve open

Inlet valve

Combustion chamber

Piston rings

Exhaust valve

Cylinder wall

Piston

Connecting rod

Figure 10.6 energy transfer in an internal combustion engine. (1) Electrical energy from the battery ignites a mixture of petrol and air in the combustion chamber: Petrol + air → carbon dioxide + water vapour + other gases + energy. (2) Chemical energy is converted to: (3) light energy, and (4) heat energy. The increased pressure due to the change in state from liquid to gas, and as hot gases expand, is (5) the mechanical energy that moves the piston down and drives the car. As a result of friction, some mechanical energy is changed to (6) heat energy

measure accurately the rate at which work is done (the rate at which energy is used, or the rate of energy transfer).

Power is *the rate of doing work*, or *the rate of energy transfer*. The SI unit for power is the watt (symbol W), named after James Watt (1736–1819), a Scot, who invented the first effective steam engine. *One watt is the rate of doing work when one joule of energy is transferred in one second.*

$$\text{Power (watts)} = \frac{\text{work done or energy transferred (joules)}}{\text{time taken (seconds)}}$$

Example. If you weigh 800 N, the work done when you step on a chair 45 cm high is $800 \times 0.45 = 360$ joules. And the rate at which you do work if you step on and off the chair 14 times in 20 seconds is:

$$\text{Power} = \frac{\text{work done}}{\text{time}} = \frac{800 \times 0.45 \times 14}{20} = 252 \text{ W}$$

Investigation 10.5   *Your rate of doing work*

You will need a strong chair, a mm rule and a stop watch.

Proceed as follows
Do not attempt this investigation if you are not fit enough. Consult your doctor if in doubt.
1 Get someone to hold the chair steady.
2 Step on and off the chair as many times as you can in 20 seconds. Count the number of step-ups.
3 Calculate your rate of doing work. Remember 1 kg = 9.8 N.

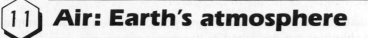

11 Air: Earth's atmosphere

11.1 Air: a mixture of gases

Earth's atmosphere is a mixture of colourless gases (see Figure 11.1), held near Earth by the pull of Earth's gravity. You cannot see air, but you feel it when the wind blows, when you move quickly through still air, and as you breathe deeply.

One colourless gas in air is water vapour. The amount in a sample of air, in a container, can be determined by introducing some anhydrous (dry) calcium chloride — which absorbs water. The increase in mass of the calcium chloride is a measure of the amount of water in the air sample. The amount of water vapour the atmosphere can hold depends on temperature. If warm air containing as much water vapour as it can hold (air saturated with water vapour) is cooled, for example by contact with a colder window, the water vapour condenses and you see water on the window.

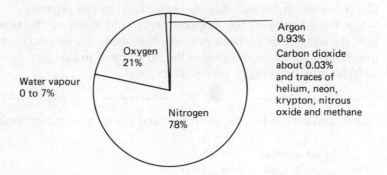

Figure 11.1 composition of dry air. Air can also contain man made gases and up to 7 per cent water vapour

When a large volume of air is bubbled through calcium hydroxide solution (lime water), the liquid becomes clouded as a suspension of calcium carbonate in water is formed:

$$Ca(OH)_2 \ (aq) \ + \ CO_2 \ (g) \longrightarrow CaCO_3 \ (s) \ + \ H_2O \ (1)$$

This indicates that a large volume of air contains a very small amount of carbon dioxide.

Note
(1) Solids that absorb water from air and yet remain solid (for example, calcium chloride) are described as hygroscopic.
(2) When as a result of a chemical reaction a solid is formed in suspension in a liquid, the solid is called a *precipitate*. It can be separated from the liquid by filtration (see page 84).

Investigation 11.1   *The oxygen content of air*

You will need two straight-sided glass jars or measuring cylinders, a bowl of water, iron filings and two pieces of flexible tubing.

Proceed as follows
1 Fill the jars with water, then empty them to make sure they contain fresh air.
2 Sprinkle iron filings over the damp inner surface of one jar.
3 Place the open end of each jar (with plastic tubes as in Figure 11.2) in the bowl of water so that air is trapped in the jars and the water level is the same inside and outside. Remove the tubes, so the water separates the air in the jars from the air outside.
4 Measure and record the height of the column of air in each jar.
5 Leave the apparatus undisturbed for a few days. Measure and record the volumes of trapped air each day. Note that the iron filings become rusty and that the water level in this jar rises.
6 When the water level has stopped rising, add water to the bowl until the water level corresponds with that inside the jar containing iron filings. Measure and record the air trapped in this jar.
7 Calculate the percentage rise in water level.

A student carried out Investigation 11.1 and made these measurements:

Air in jar at start   = 25 cm
Air in jar at end   = 20 cm
Rise in water level   = 25 − 20
Percentage rise $= \dfrac{25-20}{25} \times 100 = \dfrac{5}{25} \times 100 = 20\%$

The rise in water level in the jar containing iron filings indicates that some of the trapped air has been used. Assuming the lengths measured are proportional to the volume of gas in the jar, then the percentage rise in water level is a measure of the percentage loss of volume of this air.

The jar without iron filings is what scientists call a *control*. If the water level does not rise in this jar, you may conclude that it is the rusting of the iron filings that causes the water level to rise in the other jar, not some other change that occurs even in the absence of iron filings.

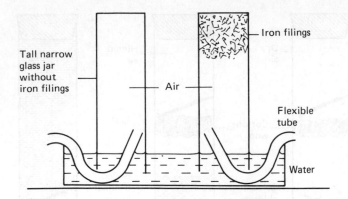

Figure 11.2 apparatus for Investigation 11.1

If a burning spill is put into the jar at the end of your investigation it will be extinguished immediately. You can conclude from this that the part of the air used when iron rusts is the same part needed for things to burn. Many investigations have been carried out in which a chemical (for example, phosphorous) has been burnt in an enclosed sample of air, and always about 20 per cent of the air has been used. The gas used in rusting and burning, which constitutes about 20 per cent of the atmosphere, is oxygen. The remaining 80 per cent of air is mainly nitrogen (see Figure 11.1).

Investigation 11.2 *Conditions needed for the rusting of iron*

You will need three test tubes with rubber or plastic bungs, 3 iron nails, dry cotton wool, anhydrous calcium chloride as a drying agent (see page 116) and distilled water that has been boiled to remove any dissolved oxygen.

Proceed as follows
1 Arrange the three tubes as in Figure 11.3
   A with anhydrous calcium chloride to absorb any water present in the trapped air
   B with distilled water and air
   C with distilled water but no air
2 A week later, examine the tubes and record your observations. Have all the nails rusted? If not, does rusting occur: (a) in air that is without water vapour, (b) in humid air, or (c) in water that is without oxygen? What can you conclude about the conditions needed for the rusting of iron?

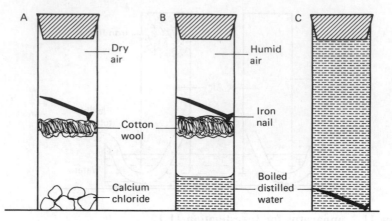

Figure 11.3 apparatus for Investigation 11.2

Iron rusting is an example of *corrosion* (wearing away from the surface by chemical action). Rusting can be minimised or prevented, for example, by smearing nuts and bolts with grease which repels water, and by painting the steel bodywork of a car to exclude water and oxygen.

Living things affect the composition of air

We breathe air throughout life. Anyone deprived of air will soon die. Clearly, we obtain something essential from the air. In 1771 a British chemist, Joseph Priestley, experimented with plants and animals in closed containers and was surprised to find that whereas an animal would soon show signs of discomfort, a green plant could live for many months.

Priestley knew that burning a candle in a closed container made the air unsuitable for either further combustion or for animal life; but he discovered that green plants restored the air's ability to support either combustion or animal life.

As a result of Priestley's investigations, and those of later scientists, on the effects of living organisms on the composition of air, we now know the explanation for Priestley's observations.

1 Oxygen is used and carbon dioxide is produced all the time by plants and animals (also when anything burns).
2 Green plants, in sunlight, use more carbon dioxide than they produce. And they produce more oxygen than they use. On balance, therefore, green plants in sunlight absorb carbon dioxide from the air and add oxygen to the air.

Investigation 11.3 *Inspired and expired air*

You will need the apparatus illustrated in Figure 11.4
Disinfect mouthpiece before and after use.

Proceed as follows
Breathe gently through the mouthpiece. You will draw air into your lungs through the lime water in tube A. As you breathe out, the expired air will pass through the lime water in tube B.

Note that the lime water in tube B turns cloudy before that in tube A, (see page 115) indicating that there is more carbon dioxide in the air you breathe out than in the air you breathe in. What do you conclude?

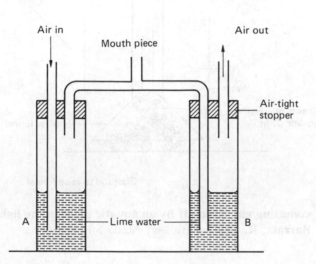

Figure 11.4 apparatus for comparing the carbon dioxide concentration of air breathed in through tube A, with that breathed out through tube B. (From Barrass, R. Human Biology Made Simple)

Investigation 11.4 *Green plants produce oxygen in sunlight*

You will need two sets of the materials illustrated in Figure 11.5

Proceed as follows
1 Place the pond weed in a beaker of pond water in sunlight, below the glass funnel and tube full of water.
2 At the same time, as a control experiment, arrange another beaker, funnel and tube with pond water but with no pond weed.

3 Note that a gas displaces the water in the test tube above the pond weed (see Figure 11.5). If any gas is lost from the pond water (the beaker with no plant present) it will be collected in the other tube.

4 When enough gas has been collected in tube A, insert a glowing splint into the tube. Oxygen relights a glowing splint, but even if the splint bursts into flame this does not prove the gas is pure oxygen. It proves only that the gas is more than 30% oxygen by volume).

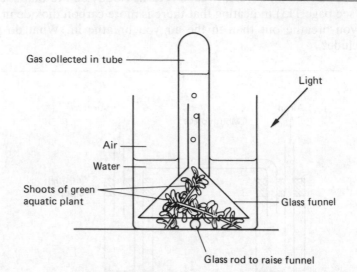

Figure 11.5 collecting gas given off by an aquatic plant in the light. (From Barrass, R. Human Biology Made Simple)

Investigation 11.4 is an experiment to test the hypothesis that green plants produce oxygen in sunlight. Was any gas collected in the control? If not, what do you conclude? What is the reason for the control experiment? Do you think the oxygen collected in the tube above the pond weed came from the plant?

Note. Other investigations on the effects of green plants on the composition of the atmosphere are included in Chapter 15.

11.2 Atmospheric pressure

Measuring air pressure

Instruments used to measure atmospheric pressure are called *barometers*. Some contain a tube of mercury (see Figure 11.6). If a tube 900 mm long is

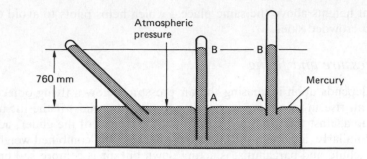

Figure 11.6 measuring atmospheric pressure. (Based on Keighley et al
   Mastering Physics)

filled with mercury and then inverted with its open end in a trough of
mercury, the level of mercury in the tube will be about 760 mm above the
level of the mercury in the trough (see Figure 11.6). Note that the level of
mercury in the tube does not change if the position of the tube is changed.
The pressure at A (in Figure 11.6) due to the column of mercury in the
tube, is equal to the atmospheric pressure at A. The vertical height AB
(760 mm in Figure 11.6) is a measure of the atmospheric pressure,
expressed in millimetres of mercury.

> **Warning.** Mercury vapour is poisonous, so mercury should be
> used only by teachers and only in conditions where nobody can
> breathe in the vapour.

The level of mercury in a barometer changes as the atmospheric pressure
changes (see weather forecasting, page 274) but *an average air pressure at
sea level is about 760 mm mercury*, which is called *one atmosphere*. At one
atmosphere, pure water boils at 100 °C. Atmospheric pressure decreases
with altitude which is why water boils at 72 °C at the summit of Mt Everest
(at 8848 m). Deep in a mine, and in a pressure cooker, water boils at above
100 °C.
   Because atmospheric pressure changes with distance above sea level and
with distance below the ground, a barometer can be adjusted for use as an
altimeter (to measure altitude). Most barometers and altimeters do not
contain mercury but they all have a scale marked in millimetres and should
give identical readings if used at the same time and in the same place.
People using different instruments, therefore, measure in the same units
and so can compare, for example: (1) barometer readings made at the
same time in different places (see page 271), or altimeter readings taken at

different heights above the same place – which helps pilots to avoid each other in crowded skies.

Air pressure and flying

Flight depends upon increasing the air pressure below a flying object. A glider can rise in ascending warm air currents (thermals) if the lift force acting up against the wings is greater than the weight of the glider, acting down. Similarly, a parachutist falls slowly because the combined weight of the parachute and parachutist is acting down but air is compressed below the canopy (see Figure 11.7C).

The fruits of some seeds resemble parachutes (see Figure 11.7B): they are held in the air and are carried in air currents. Some fruits have wings (see Figure 2.4D) that break their fall and they may be lifted and carried by the wind. This is passive non-directional flight.

An aeroplane with an engine is capable of directional flight. Air is forced in one direction, increasing air pressure behind the engines, and the aeroplane moves in the opposite direction. As the plane moves forward its wings press against the air, and because these wings are fixed at an angle (see Figure 11.7A) the higher air pressure below them, lifts the plane off the ground and keeps it in the air.

Birds have large but light-weight wing feathers, that differ from those covering the rest of the body. The wing feathers greatly increase the area of the forelimbs and form a strong but flexible surface which presses the air both back and down in flight, providing both forward drive and lift. Bats and insects also have wings and, like birds, are capable of active directional flight.

Air pressure and breathing

You have two lungs. Each one is in a pulmonary cavity which it fills almost completely. Between the lungs your heart is in a pericardial cavity, and below your heart and lungs there is a muscular diaphragm which, with the rib cage (see page 51), is used in breathing.

Your windpipe (or trachea) is a tube that opens at one end into the pharynx (see Figure 11.8) and divides at the other end into two bronchi — one of which enters each lung (see Figure 18.5). In the lung, the bronchus divides into smaller tubes or bronchioles which divide many more times. Each of the finest branches ends blindly in a bunch of air sacs. The whole of this system of air passages and air sacs contains air.

Breathing in results from muscle contraction, which raises the ribs and lowers the diaphragm (Figure 11.8A); and breathing out results from the elasticity of the lungs and the return of the ribs and diaphragm to their resting position (Figure 11.8B). These movements alternately (1) enlarge the thorax, reducing air pressure inside the lungs, with the result that air is forced into the lungs by the higher air pressure outside the body, and (2)

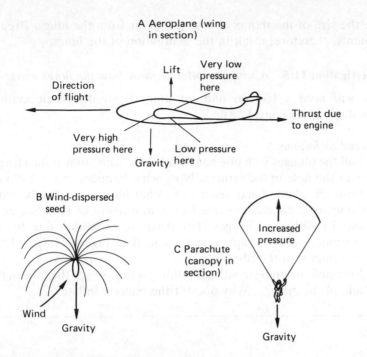

Figure 11.7 flight: canopies and wings. The velocity of an aeroplane at take off must be enough to generate a lift force greater than the force of gravity. Drawings not to scale.

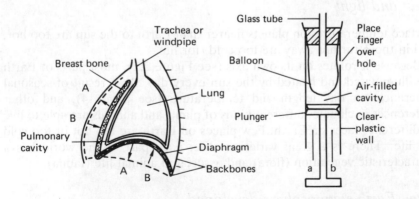

Figure 11.8 what happens when you breathe. Movement of the rib cage and diaphragm (diagram not to scale); and a simple model used to demonstrate the action of a suction pump. A = inspiration; B = expiration. (From Barrass, R. Human Biology Made Simple)

reduce the size of the thorax and so pump air from the lungs. Breathing movements, therefore, result in the ventilation of the lungs.

Investigation 11.5   *A simple model to show how the lungs work*

You will need a balloon and an unused, sterile plastic syringe (modified as in Figure 11.8).

Proceed as follows:
(1) Pull the plunger with one hand and at the same time hold a finger over the hole in the syringe. Note what happens to the balloon.
(2) Push the plunger and again note what happens to the balloon. Compare these changes due to the movement of the plunger (a and b) with the changes that occur in breathing due to the movement of the diaphragm (A and B in Figure 11.8) and the movements of the rib cage.
(3) Now pull the plunger without holding a finger over the hole in the side of the syringe. Why doesn't the balloon inflate?

11.3   Life on other planets?

People like to speculate about the possibility of life elsewhere in the universe. It may be that there is life in other solar systems but, from observations made from Earth and during space exploration, it appears that life as we know it could not exist elsewhere in this solar system.

Heat and light

Surface temperatures on planets nearer than Earth to the sun are too hot, and in those further away are too cold, for life.

Because it rotates on its own axis (see Figure 2.2), most parts of Earth are illuminated and heated by the sun every day. As a result of seasonal differences in day length and temperature (see page 14), and other differences in climate, different kinds of plants and animals are able to live in different places on Earth. Few places on Earth are too hot or too cold for life. There is a great variety of life: each part of the world has a characteristic vegetation (flora) and associated animal life (fauna).

How Earth's atmosphere developed

Jupiter is a much larger planet than Earth (see page 11). It exerts a larger gravitational pull (see page 62) which has held its original atmosphere of hydrogen and helium. Earth, being much smaller, exerts a smaller gravitational pull. This is why: (a) it lost its original atmosphere (also hydrogen and helium) and (b) its present atmosphere of heavier gases was able to develop.

Our solar system was formed about 5000 million years ago but there were no organisms on Earth for its first 1000 million years. Since then oxygen and carbon dioxide have been produced by living organisms. Indeed, the fairly constant composition of Earth's atmosphere is the result of the constant exchange of nitrogen, oxygen and carbon dioxide between living organisms and their surroundings.

Conditions on Earth were favourable for the origin of life, and living organisms have played a part in the development of Earth's atmosphere. In contrast, our moon — beause it is much smaller than Earth — exerts a much smaller gravitational pull and has no life.

Earth's atmosphere is necessary for life as we know it

1 The atmosphere is a reservoir for water vapour (see water cycle, page 271). In most places on Earth there is water, which is essential to life.
2 The atmosphere is a reservoir of the gases nitrogen, oxygen and carbon dioxide produced by living organisms and used by them.
3 The atmosphere acts like a blanket, reducing the amount of energy reaching Earth from the sun and slowing heat loss from Earth. As a result of this, and the rotation of Earth, most places are neither too hot nor too cold for life.
4 The upper atmosphere absorbs harmful particles and harmful radiations (see page 275) which, otherwise, would constantly bombard us from outer space.

11.4 Polluting Earth's atmosphere

In addition to the gases listed in Figure 11.1, very small dust particles and microbes are suspended in the atmosphere. Air also contains waste gases, produced when fuel is burned and in many industrial processes. Many of these gases, including carbon monoxide, hydrogen chloride and sulphur dioxide are harmful to people (see page 213) and to other organisms (see page 379). They are called *pollutants*.

As a result of our own activities, the atmosphere of Earth is being made less fit for life. However, people are at last realising that Earth with its atmosphere is a closed system. That is to say, the harmful waste materials which we put into the air do not disappear. They may stay in the air, or they may contaminate animals and plants and their surroundings, or they may break down (be degraded) into other materials.

Whenever practicable, potentially harmful chemicals should be degraded where they are produced instead of being released as pollutants into the air.

⬡ 12 Machines

A machine is a device that helps us to convert energy into useful work. We need to consider only simple machines.

12.1 Levers

You use a lever, for example, when you cut with scissors or open a can of paint. *A lever is a device that turns on a pivot.*

Investigation 12.1 *Using levers*

1 Close a door first by pushing near the hinge and then by pushing at the edge furthest from the hinge. Which is easier?
2 Use a spanner to loosen a nut, first holding the spanner near its middle and then by holding it as far from the nut as possible. Which is easier?

From such experiences you learn that the force required to perform a task with a lever depends upon where the force is applied. The force required decreases as the distance from the turning point or pivot increases.

The turning effect of a force, called the *moment* of the force, is calculated by *multiplying the force by the perpendicular distance of the force to its pivot* (see Figure 12.1).

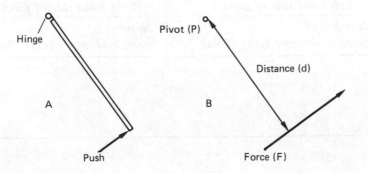

Figure 12.1 the moment of a force: (A) door and its hinge; (B) moment of force (F) about pivot (P) = F × d

Investigation 12.2   *The turning effect of a force*

You will need a 30 cm or 50 cm ruler, some metal discs (for example, identical coins) and a pivot.

Proceed as follows

1 Pivot the ruler at its mid point (as in Figure 12.2A). If necessary fix a small piece of plasticine or similar material to the lighter end of the ruler to make it rest horizontally.

2 Pile a number of coins on the left side of the ruler and a different number on the right side. Adjust the positions of the piles of coins until the ruler is again horizontal (as in Figure 12.2B). In this position the ruler is said to be in *equilibrium* (balanced).

3 Prepare a table (like Table 12.1) in your notebook. Measure the distance of each pile of coins from the pivot. Record in your table the number of coins in each pile and its distance from the pivot.

4 Calculate the moment of each force by multiplying the number of coins in each pile by its distance from the pivot. Enter your results in your table.

5 Repeat this investigation with a different number of coins in each pile and placed at different distances from the pivot.

You will find, *when two forces are acting on a body that is in equilibrium:* (1) that the moment of the force (its turning effect) is calculated by multiplying the size of the force by its distance from the pivot, and (2) *that the moment of the force on the left of the pivot is always equal to the moment of the force on the right.*

Table 12.1   *Observations (data) and results of calculations from investigation 12.2*

Left hand side of pivot		Right hand side of pivot	
Number of coins × *distance* (cm)	*Total*	*Number of coins* × *distance* (cm)	*Total*

Investigation 12.3   *Clockwise and anticlockwise moments*

You will need a half meter rule, a pivot, and a selection of known weights.

Proceed as follows

1 Place two weights on the left of the pivot, then balance them with two weights placed at different distances on the right of the pivot.
2 Prepare a table (like Table 12.2) in your notebook. Record the forces (weights), distances from pivot, and moments acting on both sides of the pivot.
3 What is the sum of the moments on the left (the anticlockwise moments) and on the right (the clockwise moments)?

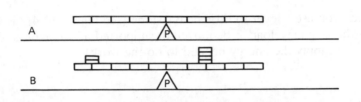

Figure 12.2 the turning effect of a force

The results of these two investigations are summarised in the *law of moments* which states that *when an object is in equilibrium, the sum of the clockwise moments equals the sum of the anticlockwise moments.*

Example. Two children are playing on a see-saw. The girl, who weighs 55 kg is sitting 2m from the pivot (see Figure 12.3). Where must the boy, who weighs 40 kg, sit to balance the see-saw?

Let d = distance of boy from pivot (P) to balance see-saw
Moments around P are: $40 \times d = 55 \times 2$

Therefore, $d = \dfrac{55 \times 2}{40} = 2.75$ m

The boy must sit on the other side of the see-saw 2.75 m from the pivot.

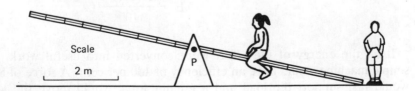

Scale

2 m

Figure 12.3 a see-saw

Table 12.2 *Observations (data) and results of calculations from investigation 12.3*

Anticlockwise moments						Clockwise moments					
Wt (N)	d 1 (mm)	moment N × mm	Wt 2 (N)	d 2 (mm)	moment N × mm	Wt 3 (N)	d 3 (mm)	moment N × mm	Wt 4 (N)	d 4 (mm)	moment N × mm
Sum of anticlockwise moments =						Sum of clockwise moments =					

When you use a lever to lift a heavy stone (see Figure 12.4), as with all levers there is: (1) a load to be moved or supported, (2) a pivot, and (3) an effort (to supply the energy needed to do the work).

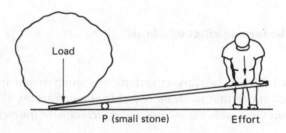

Load

P (small stone)   Effort

Figure 12.4 a lever used to lift a heavy stone

Example. When a crow-bar is used to lift a heavy stone, a force equal to the weight of the crow-bar acts through the crow-bar's centre of gravity (see page 65). However, ignoring the weight of the crowbar, apply the law of moments. If the load is 2000 N and this is 4 cm from the pivot, what effort must be applied 100 cm from the pivot to just support the load?

$$\text{Sum of anti-clockwise moments} = \text{Sum of clockwise moments}$$
$$E \times 100 = 2000 \times 4$$
$$E = \frac{2000 \times 4}{100} = 80 \text{ N}$$

If all the energy of the effort were converted into useful work, this simple machine would have an efficiency of 100 per cent. A force of 80 N would just support the load and a greater force would move the load. However, as we shall see, a machine is never 100 per cent efficient.

If in this example the pivot were now moved, to 2 cm from the load, what effort would have to be applied 100 cm from the pivot to support the same load?

$$E = \frac{2000 \times 2}{100} = 40 \text{ N}$$

You may conclude that the force needed to support a load with a lever decreases as the distance from the pivot to the load decreases.

To make your work as easy as possible, using a lever, the load must be as near to the pivot as possible and the effort must be applied as far from the pivot as possible.

Examine the drawings of some levers (Figure 12.9). When each of these levers is used, where should the load be placed to make the work as easy as possible? All these levers are made by people. However, many levers occur naturally. For example, there are levers in your body (see Figure 12.10).

12.2 Pulleys

Pulley wheels can be used to make work easier. For example, with a pulley wheel fixed at the top of a building a load can be lifted by exerting a downward force on a rope (see Figure. 12.5A). The effort moves the same

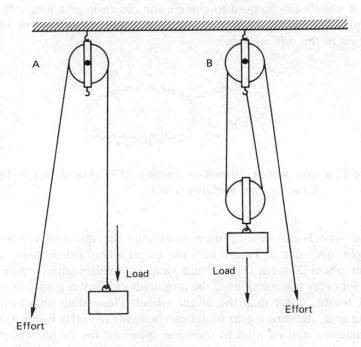

Figure 12.5 pulleys: (A) using a pulley to raise a load; (B) a double pulley system

distance as the load. In addition to the energy used in lifting the load, some energy is used in turning the pulley wheel, but the work is made easier because it is easier to pull downwards than upwards.

A combination of fixed and movable pulleys is called a block and tackle. For example, two pulley wheels (one fixed and one movable, as in Figure 12.5B) are used to make a double pulley system. This could be used in a garage to lift an engine out of a car.

With this double pulley system, the lower pulley wheel plus the load are raised 1 m when the effort moves 2 m. The effort moves twice as far as the load but a heavier load can be moved than would be possible with a single pulley. The work done by the effort is the force multiplied by the distance over which it is applied.

Potential energy is stored in the load as it is raised (see page 104). But more energy is used by the effort, because: (1) it moves twice as far as the load, (2) it raises the pulley wheel as well as the load, and (3) it must overcome friction in turning the pulley wheels.

12.3   Gears

Because a gear wheel (called the driving gear) has teeth around its edge, it can be used to turn a second gear wheel (called the driven gear). Note that the teeth on these wheels must be identical if they are to fit exactly.

Gear wheels can be used to change the direction of a force. Note, for example, that in Figure 12.6 the driven wheel rotates in the opposite direction to the driving wheel.

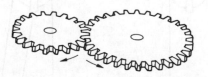

Figure 12.6 gear wheels. (Based on Dudley, D.W. Handbook of Practical Gear Design, McGraw-Hill)

Gear wheels can also be used to change the speed of rotation. For example, note that in Figure 12.6 the larger wheel has 30 teeth and the smaller wheel 20 teeth. This means that the smaller wheel rotates three times for every two rotations of the larger wheel. That is to say, the smaller wheel rotates faster than the larger wheel. Depending on which is the driving gear, the driven gear wheel can be made to rotate faster or slower. The machine can be used to increase speed (if the larger wheel is the driving wheel) or to increase force (if the smaller wheel is the driving wheel).

The driving gear wheel and the driven gear wheel of a bicycle are connected by a chain — they are not in direct contact. As a result, both gear wheels rotate in the same direction. The pedals are attached to the larger driving gear wheel. The smaller driven gear wheel rotates faster than the driving gear. The effort moves a load that is smaller than itself, but the load moves faster than the effort.

12.4 Slopes

A slope (or *inclined plane*) is a machine. It helps us convert energy into useful work. For example, it is easier to slide a load up a gradual slope than up a steep slope. And if a load is too heavy to lift, you may be able to push it up a slope.

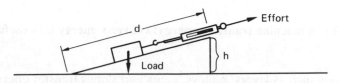

Figure 12.7 moving a load up a slope. (Based on Keighley et al Mastering Physics)

12.5 The efficiency of a machine

All machines convert energy into useful work. The energy converted is called the input of the machine; and the useful work done is called the output of the machine (see Figure 12.8).

Machines have many uses, but most machines: (1) change the direction of a force (see Figure 12.5), or (2) increase force (see Figure 12.4), or (3) increase speed (see Figure 12.6). No machine can increase force and speed at the same time.

When a load is raised, the energy input is the effort multiplied by the distance moved (E × d = J) and the energy output is the potential energy stored in the load (the load multiplied by the distance raised: L × h = J). Energy input and work done are both measured in joules (J).

┌─ **Investigation 12.4** *The efficiency of a simple machine* ──────────┐

You will need a slope, a ruler, a spring balance, and a suitable load (for example, a block of wood or a brick) and string. See Figure 12.7.

Proceed as follows
1 Use the string to fasten the load to the spring balance.

2 Measure the length and the height of the slope, and the force needed to move the load. Record your data in your notebook.
3 Calculate the efficiency of this machine:

$$\text{Efficiency of machine (\%)} = \frac{\text{Load} \times \text{h}}{\text{Effort} \times \text{d}} \times 100$$

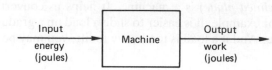

Figure 12.8 a machine transfers energy: converts energy into useful work

Machines such as levers, pulleys, gears and slopes transfer energy from the effort to the load, but some mechanical energy is always transferred to heat energy due to friction. When a lubricant is used to reduce friction, it reduces wear between surfaces and increases the efficiency of the machine.

The efficiency of a machine is a measure of how much energy is transferred in an intended way – the useful energy output or useful work done expressed as a percentage of the energy input.

$$\text{Efficiency} = \frac{\text{energy transferred as desired}}{\text{total energy transferred}} \times 100$$

Energy can be neither created nor destroyed (see page 104), so the output of a machine can never be greater than the input. Indeed, because of friction, the energy used in doing useful work is always less than the energy input. No machine is 100 per cent efficient.

If the load raised by a machine is greater than the effort, then the distance the load moves must be less than the distance the effort moves (see Figure 12.4, Figure 12.5B and Figure 12.7).

If the load moves further than the effort, the effort applied must be greater than the load (see Figure 12.11C).

The machines designed by mechanical engineers, working to scientific principles, include wheels, axles, pulleys, pivots and levers. Some everyday examples of levers, for example, are shown in Figure 12.9. The same scientific principles apply to the bones and muscles that support your body, and to the bones, joints and muscles used when one part of your body moves in relation to another part.

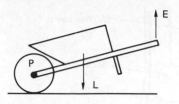
Wheelbarrow

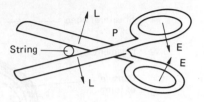

Scissors

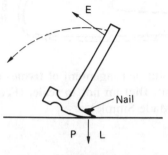
Claw hammer

Figure 12.9 levers in everyday use

12.6 Posture and movement

Your body is supported by your skin, your skeleton, and the muscles that pull upon your skeleton — enabling you to move from place to place, move one part of your body in relation to another, or just keep still. That is to say, your skeleton and skeletal muscles support your body, and make possible both movement and the maintenance of posture.

Your skeleton and skeletal muscles, together, enable your body to resist most of the pushes and pulls (the forces) to which it is subjected. In standing upright, for example, your body resists the force of gravity. If your skeletal muscles were to stop holding your bones in place, as in a faint, you would collapse.

Joints and movement

In your arm, the ulna moves upon the humerus at the elbow (see Figure 12.10). The elbow is a hinge joint. Pads of cartilage at the ends of the bones act as shock absorbers, a lubricating fluid reduces friction between the cartilage pads, and a fibrous joint capsule holds the other parts in place.

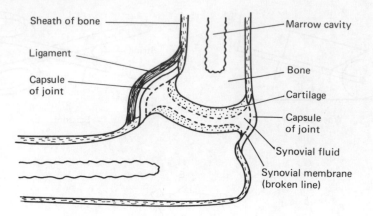

Figure 12.10 a hinge joint: arrangement of tissues in a section through an elbow joint. Digram not to scale. (From Barrass, R. Human Biology Made Simple)

Investigation 12.5   *Looking at a joint*

You will need a fresh joint (from, for example, a sheep or a rabbit). Wash your hands thoroughly and wipe work area with disinfectant.

Proceed as follows
Study the arrangement of the bones, cartilages and ligaments. Find out how one bone moves in relation to the other. Muscles are attached to bones, near to joints, by inelastic tendons.

Your biceps muscle bends (flexes) and your triceps muscle straightens (extends) your forearm. That is to say, the force exerted by the contraction of one muscle acts in the opposite direction to the force exerted by the other muscle. They work as a mutually antagonistic pair.

Bones as levers

Depending on the position of attachment of a muscle to a bone, in relation to a joint, a lever either increases the movement caused by muscle contraction or increases the force generated. Like any other machine, it cannot do both.

When you lift a weight placed on your hand, by bending your arm, the small movement of the ulna near to the joint — where the muscle is attached — moves the hand much further. The effort is applied between the pivot and the load (see Figure 12.11C), so a large upward force (the effort) is needed to balance a smaller downward force (the load).

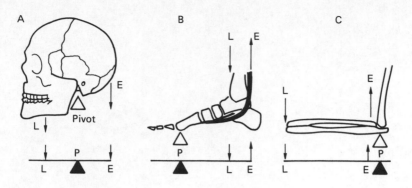

Figure 12.11 bones as levers. Diagram not to scale. (From Barrass, R. Human Biology Made Simple)

When you stand on tiptoe the load is between the effort and the pivot (see Figure 12.11B), as with a wheelbarrow. The effort of muscle contraction is amplified and lifts the weight of your body, overcoming a larger force. But note that the heel, where the effort is applied, moves further than the load.

Your head is balanced at the end of your vertebral column. This is a balanced lever system, like a see-saw. The load (the weight of the part of your head in front of the pivot) is balanced partly by the weight of your head behind the pivot and partly by the effort of muscle contraction. Very little effort is needed when you stand erect or sit upright (see Figure 12.11A).

Good posture

An upright posture is a good posture. Your weight is evenly distributed and you can, therefore, maintain the posture with the minimum effort. Breathing deeply causes you to adopt an upright posture, and the upright posture makes breathing easier. If your posture is good you feel good. People are unique in standing and walking erect. This upright posture is adopted naturally by young men and women who are in good health, physically fit and mentally alert.

13   **The structure of matter**

13.1 The parts of an atom

In the 1890s scientists in many countries were investigating the cathode rays produced when a high–voltage electric current was passed through a glass tube from which all the air had been removed (a vacuum). In 1897, Joseph Thomson, a British scientist, proved that these rays were streams of negatively charged particles (now called electrons). They were given off by the heated cathode (see Figure 13.1) and attracted to the anode, and were therefore called cathode rays. Some of them, passing through a hole in the anode, caused a green glow on the wall of the tube.

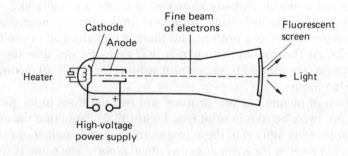

Figure 13.1 a cathode ray tube. (Simplified from Keighley et al Mastering
         Physics)

Electrons, because they are electrically charged particles, are deflected by an electromagnetic field. This fact is now used in a cathode ray oscilloscope to measure the voltage of an electric power supply and to display variations in voltage (for example, see Figure 32.3A).

Thomson discovered that a hydrogen atom, which is the smallest atom (see page 87), has a mass 1837 times greater than that of one of these particles. So, as a result of these investigations by physicists, chemists had to accept that the hydrogen atom could no longer be regarded as the smallest particle possible.

The particle theory of matter (see page 39) has had to be modified in the light of these observations and the more recent work of other scientists. By 1908 Rutherford, another British scientist, had established that most of an atom's volume was occupied by electrons and that the atom had a much denser core. Because the electron was negatively charged, and the atom as a whole was electrically neutral, it was at first thought that the core, or atomic nucleus, contained only positively charged particles (called protons).

But in 1930, two German physicists, Bothe and Becker, discovered a new radiation from the atomic nucleus. And another British physicist, Chadwick, showed that this radiation was composed of neutral particles (neutrons), with no electrical charge and each with a mass equal to that of one proton.

> Each *electron* has a single negative electrical charge.
> Each *proton* has a single positive electrical charge.
> Each *neutron* is electrically neutral: it has no electrical charge.

The number of protons in the nucleus is exactly balanced by an equal number of electrons in orbit, so the atom as a whole has no net electrical charge: it is electrically neutral.

In 1936, a Danish physicist, Niels Bohr, suggested a simple model which will help you to think about atomic structure (see Figure 13.2). Remember that this is just a model. Nobody knows what atoms are really like.

The smallest atom, hydrogen, is unique in having no neutrons. Its nucleus comprises only one proton, and there is one electron in orbit (see Figure 13.2). As the mass of hydrogen is 1837 times greater than the mass of one electron (see page 139), the proton acounts for nearly all the mass of the hydrogen atom.

Each atom of helium has two protons and two neutrons in its nucleus, and there are two electrons in orbit (see Figure 13.2). Note that the *atomic number* (or proton number) of these two elements is *the number of protons or electrons in each of the element's constituent atoms*. The same is true of

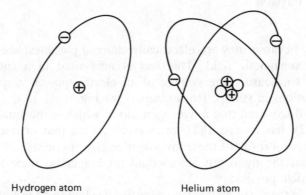

Hydrogen atom   Helium atom

Figure 13.2 Bohr model of atomic structure. Key: ⊖ = electron, ⊕ = proton, ○ = neutron. Each line represents the orbit of one electron

all other elements. The atoms of each element in the periodic classification have one more proton and one more electron than the atoms of the preceding element in the series.

13.2 How atoms combine

Protons and neutrons are of great scientific interest and economic importance (see page 330), but the chemical combination of atoms depends only on the electrons.

The electrons in orbit around the nucleus of an atom are in layers. The first layer, nearest the nucleus, can contain one or two electrons. The second layer can contain up to eight electrons. The number of electrons in orbit in the outer layer of any atom determines the atom's chemical properties; that is to say, its ability to combine with other atoms to form molecules or giant structures. Note the number of electrons in the outer layers of atoms of elements 1 to 20 of the periodic tables (see Table 13.1).

Table 13.1   *Arrangement of electrons in atoms 1 to 20 of periodic table*

Element	Symbol	Atomic number	Layers of electrons in atom			
			1	2	3	4
Hydrogen	H	1	1			
Helium	He	2	2	*First layer complete*		
Lithium	Li	3	2	1		
Beryllium	Be	4	2	2		
Boron	B	5	2	3		
Carbon	C	6	2	4		
Nitrogen	N	7	2	5		
Oxygen	O	8	2	6		
Fluorine	F	9	2	7		
Neon	Ne	10	2	8	*Second layer complete*	
Sodium	Na	11	2	8	1	
Magnesium	Mg	12	2	8	2	
Aluminium	Al	13	2	8	3	
Silicon	Si	14	2	8	4	
Phosphorus	P	15	2	8	5	
Sulphur	S	16	2	8	6	
Chlorine	Cl	17	2	8	7	
Argon	Ar	18	2	8	8 *Third layer complete*	
Potassium	K	19	2	8	8	1
Calcium	Ca	20	2	8	8	2

Note that group 0 of the periodic table (see page 87) includes the elements helium, neon and argon. The outer layer of each of these atoms is complete (with 2, 8 and 8 electrons, respectively, see Table 13.1) The atoms of these elements are so stable that they cannot combine with other atoms: they exist as separate atoms, not as molecules. They are all gases which, *because they are unreactive*, are called *inert gases*.

The metals in group 1 of the periodic classification include lithium, sodium and potassium (see Table 8.1). The outer layer of each of these atoms contains just one electron (see Table 13.1). These elements have similar physical properties: they are all soft, shiny metals. They also have similar chemical properties. For example, they all react with water, forming hydrogen gas:

$$2Na \text{ (s)} \quad + \quad 2H_2O \text{ (1)} \quad \longrightarrow \quad 2NaOH \text{ (aq)} \quad + \quad H_2 \text{ (g)}$$

sodium	water	sodium hydroxide	hydrogen
(solid)	(liquid)	(aqueous solution)	(gas)

Note the progressive difference in the reaction of these metals with water: lithium steadily, sodium vigorously, and potassium violently.

Another similarity in the chemical properties of these group 1 metals is that they all ignite spontaneously in air and burn with a brilliant flame. For example:

$$4Li \text{ (s)} \quad + \quad O_2 \text{ (g)} \quad \longrightarrow \quad 2Li_2O \text{ (s)}$$

lithium	oxygen	lithium oxide

Elements in group 7 of the periodic classification (see page 87) include fluorine, chlorine, bromine and iodine. The outer layer of each of these elements contains seven electrons (see fluorine and chlorine in Table 13.1). Note the progressive difference in the physical properties of these elements: fluorine (a pale-yellow gas), chlorine (a yellow-green gas), bromine (a red-brown liquid), and iodine (a volatile black solid). These group 7 elements also have similar chemical properties. For example, they all react with metals to form similar solids:

$$2Fe \text{ (s)} \quad + \quad 3Cl_2 \text{ (g)} \quad \longrightarrow \quad 2FeCl_3 \text{ (s)}$$

hot iron	chlorine	iron (III) chloride

$$Zn \text{ (s)} \quad + \quad Br_2 \text{ (g)} \quad \longrightarrow \quad ZnBr_2 \text{ (s)}$$

zinc wool	bromine vapour	zinc bromide

Group 7 elements all react with hydrogen to form similar gases:

$$H_2 \text{ (g)} \quad + \quad Cl_2 \text{ (g)} \quad \xrightarrow[\text{ultraviolet radiation}]{\text{explodes in}} \quad 2HCl \text{ (g)}$$

hydrogen	chlorine		hydrogen chloride

H_2 (g)	+	Br_2 (g)	reacts	$2HBr$ (g)
hydrogen		bromine	$\xrightarrow{\quad}$ slowly	hydrogen bromide

Mendeleev placed elements in order according to their atomic numbers, and in groups according to their physical and chemical properties. He left gaps in his periodic classification for elements that had still to be discovered and he predicted some of their properties (see page 87).

Mendeleev's periodic arrangement of elements provided a basis for the further development of the atomic theory (see page 39). As a result we can relate the properties of each element to the unique structure of its constituent atoms. Now we can consider why all elements within each group of the periodic table have similar chemical properties.

Atoms of the inert gases, with complete outer layers containing the maximum number of electrons, are very stable (unreactive). Clearly, a full outer layer of electrons results in a very stable atom. When the atoms of other elements combine, they do so in such a way as to fill their outer layers. By *chemical bonding* they achieve stability.

Ionic or electrovalent bonding

Sodium, an extremely reactive metal, has only one electron in its outer layer (see Table 13.1). Chlorine, a suffocating gas, has seven electrons in its outer layer (see Table 13.1): one more electron would complete its outer layer. When sodium is dropped into chlorine gas there is a violent reaction and sodium chloride (common edible salt) is formed.

In the formation of sodium chloride, one electron from the sodium atom moves into the outer layer of the chlorine atom. Each atom then has a complete outer layer (see Figure 13.3). The one that achieves stability by losing an electron is said to have been oxidised. The other, gaining an electron, is said to have been reduced.

When the sodium atom loses an electron it is left with ten electrons, but it still has 11 protons. It therefore has a net positive charge $(+1)$ and is called a sodium ion. The chlorine atom, having gained an electron, has 18 electrons and 17 protons. It has a net negative charge (-1) and is called a chlorine ion.

An ion is an atom that has lost or gained electrons and so become positively or negatively charged. The sodium (Na^+) and chloride (Cl^-) ions are held together in sodium chloride (NaCl) because they have opposite electrical charges (opposite charges attract). This is an example of ionic or electrovalent bonding. Note, however, that there is no such thing as a sodium chloride molecule. The sodium and chloride ions present in a crystal of salt are not separate sodium chloride molecules but part of a *giant structure* containing equal numbers of sodium and chloride ions (see Figure 13.3).

Na⁺	Cl⁻	Na⁺	Cl⁻	Na⁺	Cl⁻	Na⁺
Cl⁻	Na⁺	Cl⁻	Na⁺	Cl⁻	Na⁺	Cl⁻
Na⁺	Cl⁻	Na⁺	Cl⁻	Na⁺	Cl⁻	Na⁺
Cl⁻	Na⁺	Cl⁻	Na⁺	Cl⁻	Na⁺	Cl⁻
Na⁺	Cl⁻	Na⁺	Cl⁻	Na⁺	Cl⁻	Na⁺
Cl⁻	Na⁺	Cl⁻	Na⁺	Cl⁻	Na⁺	Cl⁻
Na⁺	Cl⁻	Na⁺	Cl⁻	Na⁺	Cl⁻	Na⁺

Figure 13.3 continuous ionic bonding in sodium chloride (common salt). (From Critchlow, P. Mastering Chemistry)

Ionic bonding occurs when the atoms of metals that have few electrons to lose, combine with those non-metals that need few electrons to complete their outer layers. The attraction is most vigorous between atoms of group 1 metals and atoms of group 7 non-metals.

Covalent bonding

The atoms of non-metallic elements to the right of the periodic table, which in ionic or electrovalent bonding achieve stability by gaining electrons from metallic elements, can also achieve stability by sharing electrons with the atoms of non-metallic elements. This is called covalent bonding.

For example, hydrogen and chlorine react explosively when exposed to ultraviolet radiation (see page 142). Hydrogen chloride gas is formed. If the electrons in the outer layers of the hydrogen and chlorine atoms are represented as dots and crosses, this will help you understand how they combine. They complete their outer layers of electrons, and so achieve stability, by sharing electrons.

The hydrogen atom now has two electrons in its outer electron layer (like the inert gas helium), and the chlorine atom now has eight electrons in its outer layer (like the inert gas neon).

In hydrogen (H_2), oxygen (O_2) and chlorine (Cl_2) gases, pairs of atoms are held together by covalent bonding. The outer layers in the atoms forming these molecules can be represented diagrammatically, in pairs, with dots to represent the electrons of one atom and crosses to represent those of the other atom.

Alternatively, each covalent bond can be represented by a straight line.

H — H	O = O	Cl — Cl
covalent	double	covalent
bond	covalent	bond
	bond	

Similarly, water (H_2O) and carbon dioxide (CO_2) are formed by covalent bonding.

$$H \overset{x}{\underset{\bullet}{:}} \overset{xx}{\underset{xx}{O}} \overset{\bullet}{\underset{x}{:}} H \qquad\qquad \overset{x}{\underset{x}{{}^x_x O}} \overset{\bullet}{{}^x_\bullet :} \overset{}{{}_x^x C {}_x^x} \overset{\bullet}{: {}^\bullet_\bullet} \overset{x}{\underset{x}{O {}^x_x}}$$

Alternatively: H — O — H   O = C = O

The attraction between the molecules of most covalent compounds is weak. As a result, most are gases, or liquids with low boiling points.

The valency or bonding ability of atoms of different elements

The number of electrons an atom has to gain or lose to make a stable outer layer is called its bonding ability or *valency* of the element. For example, sodium has to lose an electron (has a valency of +1) and chlorine has to gain an electron (has a valency of −1). Other valencies are included in Table 13.2.

You can use this table to help you predict the number of atoms of each element that will combine to form one molecule of a compound. For example, one atom of hydrogen (valency +1) will combine with one atom of chlorine (valency −1) to form one covalent molecule of hydrogen chloride gas (formula HCl), but for every atom of zinc (valency +2) there are two atoms of chlorine (valency −1) in the electrovalent compound zinc chloride (formula $ZnCl_2$).

Note that the atoms of metals (on the left of the periodic table) lose electrons to achieve stability. The atoms of non-metals (groups 5, 6 and 7 of the periodic table), gain or share electrons to achieve stability. Also note that carbon, and other elements in group 4, with a valency of 4, can achieve stability either by gaining 4 electrons or by losing 4 electrons.

Note also that some elements have more than one possible valency. For example, iron can lose two electrons (valency +2) and form one group of compounds, or it can lose three electrons (valency +3) and form a different group of compounds. To indicate which ion is formed the valency is shown in brackets after the name of the element, as part of the name of the compound formed. For example, iron (II) chloride has the formula $FeCl_2$ whereas iron (III) chloride has the formula $FeCl_3$.

Table 13.2   *The periodic table: valency of elements in different groups*

Groups	1	2				3	4	5	6	7	0
Valency	+1	+2	+2 +3	+1 +2	+2	+3	+4 −4	+5 −3	−2	−1	

Level 1	H		10 transition								He
Level 2	Li	Be	metals			B	C	N	O	F	Ne
Level 3	Na	Mg	including:			Al	Si	P	S	Cl	Ar
Level 4	K	Ca	Fe	Cu	Zn					Br	Kr
Level 5										I	Xe

Have more electrons than are needed.
Lose electrons to achieve stability.
Oxidised easily.

Have fewer electrons than are needed.
Gain electrons to achieve stability.
Reduced easily.

13.3   Making compounds

Compounds are chemical substances that contain more than one kind of atom, chemically combined (see page 88). It is possible to make compounds from their constituent elements. For example, iron (II) sulphide can be made from iron and sulphur (see page 108).

$$\text{iron} + \text{sulphur} \longrightarrow \text{iron (II) sulphide}$$
$$\text{Fe (s)} + \text{S (s)} \longrightarrow \text{FeS (s = solid)}$$

Iron sulphide, like sodium chloride (see page 144) is not composed of molecules. The iron and sulphide ions, attracted by their opposite electrical charges, are part of a crystal (a giant structure).

Most chemical compounds are made not by the direct combination of their constituent elements, but in a chemical reaction between compounds that contain the necessary elements. Acids and bases are two kinds of chemical substances used in the production of other chemicals.

The acids most commonly used in chemistry laboratories, for making other compounds, are hydrochloric acid (HCl), sulphuric acid (H_2SO_4) and nitric acid (HNO_3).

Acids are formed in the reaction of a non-metallic oxide with water. For example:

sulphur dioxide +   water $\longrightarrow$ sulphuric acid
SO_2 (g = gas) + H_2O (l = liquid) $\longrightarrow$ H_2SO_4 (aq = aqueous solution)

carbon dioxide +   water $\longrightarrow$ carbonic acid
CO_2 (g)   + H_2O (l) $\longrightarrow$ H_2CO_3 (aq)

All acids contain hydrogen, all or part of which can be replaced by a metal in a chemical reaction. For example:

zinc + hydrochloric acid $\longrightarrow$ zinc(II)chloride + hydrogen
Zn (s) +   2HCl (aq) $\longrightarrow$ $ZnCl_2$ (s)   + H_2 (g)

All acids dissociate when dissolved in water. This means that their constituent ions separate – called *ionisation*. For example:

nitric acid + water $\longrightarrow$ hydrogen ion + nitrate ion
HNO_3 +   aq $\longrightarrow$ H^+ (aq)   + NO_3^- (aq)

The concentration of hydrogen ions is a measure of the acidity of a solution (the strength of the acid, see page 149).

Bases are metal oxides and metal hydroxides, for example, copper oxide and sodium hydroxide. Most bases are insoluble in water. Bases that do dissolve in water dissociate, releasing hydroxyl ions (OH^-), are called alkalis. For example, sodium hydroxide dissociates in water:

sodium hydroxide + water $\longrightarrow$ sodium ion + hydroxyl ion
NaOH (s)   +   aq $\longrightarrow$ Na^+(aq)   + OH^-(aq)

The concentration of hydroxyl ions is a measure of the alkalinity of a solution (the strength of the alkali, see page 149).

Bases can be used to neutralise acids. For example:

hydrochloric acid + copper oxide $\longrightarrow$ copper chloride + water
2HCl (aq)   +   CuO (s) $\longrightarrow$ $CuCl_2$ (aq) + H_2O (l)
ionic   ionic   covalent

nitric acid + sodium hydroxide $\longrightarrow$ sodium nitrate + water
HNO_3 (aq) +   NaOH (aq) $\longrightarrow$ $NaNO_3$ (aq) + H_2O (l)
ionic   ionic   ionic   covalent

Note

1 A neutral solution is one that contains neither H^+ nor OH^- ions.
2 The reaction between an acid and a base always produces a salt and water:

$$\text{acid} + \text{base} \longrightarrow \text{salt} + \text{water}$$

3 In the formation of a salt the hydrogen of an acid is replaced by a metal. Common edible salt, sodium chloride, can be made:

hydrochloric acid + sodium hydroxide $\longrightarrow$ sodium + water
                                                          chloride

HCl (aq) + NaOH (aq) $\longrightarrow$ NaCl (aq) + H_2O (l)
ionic   ionic   ionic   covalent

Table 13.3 *Some common radicals*

	Radical		Acid formed with hydrogen ions	
Name	*Formula*	*Valency*	*Name*	*Formula*
Nitrate	NO_3^-	-1	Nitric acid	HNO_3
Bicarbonate	HCO_3^-	-1	Carbonic acid	H_2CO_3
Bisulphate	HSO_4^-	-1	Sulphuric acid	H_2SO_4
Carbonate	CO_3^{2-}	-2	Carbonic acid	H_2CO_3
Sulphate	SO_4^{2-}	-2	Sulphuric acid	H_2SO_4
Phosphate	PO_4^{3-}	-3	Phosphoric acid	H_3PO_4
Hydroxide	OH^-	-1	Not an acid radical	
Ammonium	NH_4^+	$+1$	Not an acid radical	

Some elements are commonly associated in ionic (electrovalent) compounds (see Table 13.3). These associations of elements are called *radicals*. They have valencies and exist only combined with other radicals or ions in ionic compounds or aqueous solution. For example, many radicals combine with hydrogen ions, forming acids. These are called acid radicals.

In aqueous solution radicals behave as ions — they are electrically charged. As stated on page 147 it is hydroxyl (OH^-) ions that make an aqueous solution alkaline. In many chemical reactions (see page 147) a radical present as part of a reactant is also present, unchanged, as part of one of the products.

Measuring acidity and alkalinity

The greater the concentration of hydrogen ions the stronger the acid. Hydrogen ion concentration is expressed on a pH scale (see Figure 13.4). Pure water is neutral (neither acid nor alkaline). Its pH is 7. Between pH 7 and pH 0 is acid: the lower the pH the stronger the acid. Between pH 7 and pH 14 is alkaline: the higher the pH the stronger the alkali.

The measurement of pH is important in many occupations, as well as chemistry. For example, farmers and gardeners measure the pH of soil because different plants grow best in soils of a particular pH. Bakers measure pH because a low pH gives a cake of fine texture and a high pH a cake of coarse texture. Baking soda and lemon juice are used in baking.

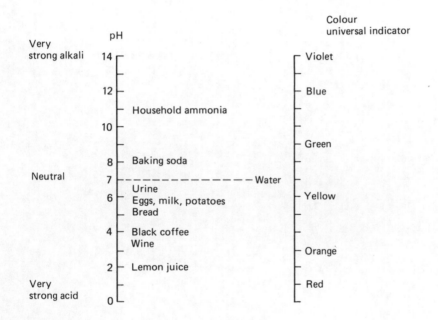

Figure 13.4 **the pH of some common liquids**

The pH of a solution can be measured by using a universal indicator. This is a mixture of dyes, available as a solution or as a test paper, which changes colour (see Figure 13.4), indicating the pH of a solution to which it is added. Litmus, another indicator, is a blue dye that changes to red if added to an acid solution. For more precise measurements a pH meter can be used. This gives a direct reading on a scale.

Warning Acids and bases are corrosive. Wear eye protection when handling acids and bases. They attack clothing and flesh and should be handled with great care. Any splashes should be washed away immediately with plenty of water.

When diluting an acid with water, always add the acid slowly to the water, stirring all the time. *Never add water to an acid*. To do so would cause dangerous spitting.

14 Static and current electricity

14.1 Static electricity

Most objects are electrostatically neutral, because the atoms of which they are composed have equal numbers of positively and negatively charged particles (protons and electrons, see page 140). Rubbing two different materials together, as in Investigation 10.4 (see page 111), does not make charged particles. It merely separates existing negatively charged particles (electrons) from the surface of one object and transfers them to the surface of the other object. As a result of losing electrons one object has more protons than electrons, and so is positively charged. At the same time the other object gains electrons and so becomes negatively charged.

Investigation 14.1 *Electrostatic attraction and repulsion*

You will need 2 polythene rods (opaque plastic), 2 cellulose acetate rods (clear plastic), a wool cloth and some thread. All items must be dry.

Proceed as follows
1 Charge one polythene rod by rubbing it with the cloth, then suspend it as shown in Figure 14.1
2 Bring the other, uncharged polythene rod near one end of the suspended rod. What happens?
3 Now charge the second polythene rod, and then see what happens when you hold it near the suspended rod.
4 What happens when you hold first an uncharged and then a charged cellulose acetate rod near one end of the suspended polythene rod?
5 What happens when you hold the wool cloth, used in charging, near the charged suspended rod? (With this simple apparatus you may not be able to demonstrate that the cloth as well as the rod is charged by friction.)
6 Repeat steps 1–5 with a suspended cellulose acetate rod.

Compare your notes, made during Investigation 14.1, with those shown in Figure 14.1. You may come to these conclusions:

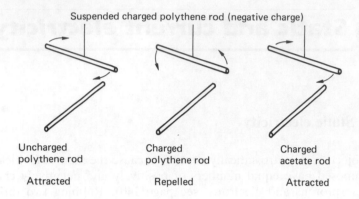

Suspended charged polythene rod (negative charge)

Uncharged polythene rod	Charged polythene rod	Charged acetate rod
Attracted	Repelled	Attracted

Figure 14.1 like charges repel, unlike charges attract

1 Some materials, including polythene and cellulose acetate, can be electrically charged.
2 There are two kinds of charge. Note that in Figure 14.1 the polythene rod is negatively charged and the cellulose acetate rod positively charged.
3 *Like charges repel.*
4 *Unlike charges attract.*

The pieces of paper used in Investigation 10.4 (page 111) were electrically neutral (like the uncharged rod in Figure 14.1). When you brought negatively charged materials near them, like charges on the pieces of paper (electrons) were repelled — leaving opposite charges close together. This explains the attraction of the pieces of paper to some charged objects and the attraction of the uncharged polythene rod to the charged polythene rod (Figure 14.1) The charged polythene rod must be negatively charged. A positively charged rod would not attract an uncharged rod, because the protons of the uncharged rod are not free to move (see page 307). Now you know how to tell if a charged material is positively or negatively charged — only negatively charged materials attract uncharged materials.

If you hold an uncharged polythene rod near a suspended negatively charged polythene rod, electrons on the uncharged polythene rod will be repelled. They will move to the end of the rod furthest from the suspended negatively charged rod – and will then be conducted through your body to earth. As a result, you will be holding a positively charged polythene rod. We say that this rod has been charged by *induction*, not by friction.

It is possible to charge the surface of a metal object by friction, but the charge is conducted through the metal and through your body to the ground. The charge is said to leak to earth. All metals and carbon are *conductors of electricity*. Most other charged materials will hold an electric charge because they do not conduct electricity. These non-conductors of electricity can be used to stop electrical charges in a conductor leaking to earth. They are, therefore, called *electrical insulators*. All plastics are good insulators.

A metal object can be charged by friction but it will hold the charge only if supported by an insulator both while it is being charged and afterwards. The conductor then holds the charge because there is no path by which electricity (moving electrons) can leak to earth.

14.2 Current electricity

The work of two Italian scientists, Luigi Galvani and Alessandro Volta, made possible the development of current electricity. In 1780 Galvani was demonstrating that a muscle in a frog's leg could be made to twitch by connecting it, using a wire (a conductor of electricity), to a machine that produced static electricity. You will remember that static electricity is the same as the electricity that flows along a wire (see page 112). On one occasion Galvani noticed the muscle twitch when it was not connected to the machine. He thought, at first, that the leg must be producing electricity. However, after many careful experiments he proved that it was not. In his experiments with frogs' legs, Galvani was using a steel knife, a copper dish, and a salt solution. He demonstrated that electricity was produced by the combined action of these materials even in the absence of a frog's leg.

Galvani's discovery of a method of producing electricity could be said to have been made by accident during an investigation designed for another purpose. This is perhaps true of all discoveries. We can plan investigations only on the basis of what we already know, so the unexpected can be discovered only by accident. Discoveries are made by people who observe carefully and are prepared to consider unexpected as well as expected observations.

Electric cells

Volta built on the foundation laid by Galvani. He devised a detector of electricity so that work on electricity could be separated from the study of muscle contraction. In 1800 he showed that when two different metals (called *electrodes*) were placed in a solution that conducted electricity (called an *electrolyte*), electricity was produced. He called this device for producing electricity an electric cell.

Volta was the first to show how chemical energy could be changed to electrical energy.

A cell that produces an electric current as a result of a chemical reaction within the cell is called a primary cell.

Investigation 14.2 *Making a primary cell*

You will need a 1.5 V bulb to act as a current detector, some connecting wire, a copper plate and a zinc plate (as electrodes), a glass jar, enough dilute sulphuric acid (an electrolyte) to half fill the jar and eye protection. See Figure 14.2.

   Warning (1) Sulphuric acid is corrosive. Do not spill it or get any on your skin. (2) Remember when diluting any acid, you must add the acid a little at a time to the water: **never add water to an acid**.

Proceed as follows
1 Pour the dilute acid into the jar.
2 Put the copper and zinc plates in the jar, keeping them apart.
3 Connect the bulb to the plates (as in Figure 14.2). **Note** that the bulb lights for a few seconds and then goes out. At the same time, bubbles of a gas form on the copper plate.
4 Remove the copper plate. Wash it under a tap. Repeat steps 2–4

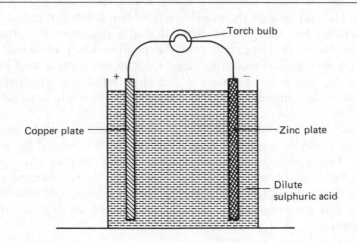

Figure 14.2 a voltaic cell (one kind of primary cell)

Chemical reactions in a primary cell

Zinc atoms give up electrons and move into the surrounding electrolyte as zinc ions:

$$\underset{\text{Zinc metal}}{\text{Zn}} \longrightarrow \underset{\text{Zinc ion}}{\text{Zn}^{2+}} + \underset{\text{Two electrons}}{2\text{e}^-}$$

This leaves the zinc metal with extra electrons, so it has a negative charge.

The zinc plate is therefore called the *negative electrode* or *cathode*.

Sulphuric acid in solution in water ionises:

$$H_2SO_4 \longrightarrow 2H^+ + SO_4^{2-}$$

Sulphuric acid   Hydrogen ions   Sulphate ion

The positively charged hydrogen ions, repelled from the zinc by the positively charged zinc ions (remember that like charges repel), move to the copper plate (remember that unlike charges attract). Here they remove electrons from the copper, producing hydrogen molecules:

$$2H^+ + 2e^- \longrightarrow H_2$$

Two hydrogen ions   Two electrons   One hydrogen molecule

The copper, having lost two electrons, has a positive charge. The copper plate is therefore called the *positive electrode* or *anode*.

The accumulation of hydrogen bubbles on the anode separates the electrolyte from the copper and stops the action of the cell. The cell is then said to be *polarised*.

Copper and zinc are not the only metals that can be used in making a primary cell. Metals can be listed in order (arranged in a series) so that when any two are used as electrodes in making a cell, the metal earlier in the list is the anode (for example: copper, lead, iron, zinc, aluminium and magnesium). This arrangement of metals is called the *electrochemical series*.

The force driving electrons from the negative electrodes, where there is a surplus of electrons, through the wire and light bulb to the positive electrode, where there is a deficiency of electrons, is called the electro-motive force (e.m.f.).

Bubbles of hydrogen gas, which cause polarisation, form on the copper plate only when the cell is working and electricity is flowing through the light bulb. If ordinary commercial zinc is used for the negative electrode, bubbles of gas come from this zinc plate all the time (even when no current is flowing). This is the result of what is called local action, caused by impurities in the zinc forming many small cells within the zinc. To give a useable source of electricity, a cell must contain both a depolariser, to remove hydrogen bubbles from the positive electrode, and a means of preventing local action.

Most primary cells used today, for example in torches and portable radios, are called dry cells. They are not completely dry. The electrolyte, ammonium chloride, is a paste (see Figure 14.3). The zinc case, which is the negative electrode, dissolves, slowly. Eventually the electrolyte seeps out, so it is best to remove dry cells from any appliance that is not in use.

If you use a dry cell for too long, or try to take too much current, it will stop working. This is because the manganese dioxide, which is used as a

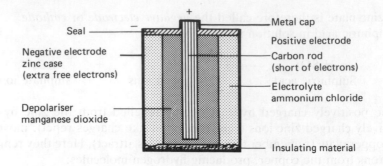

Figure 14.3 a dry cell (a type of primary cell in common use)

depolariser (providing oxygen that combines with hydrogen to form water) is not acting quickly enough to prevent depolarisation.

Another type of electric cell, used for example in a car battery, is called a secondary cell. Whereas a primary cell produces electricity as a result of an irreversible chemical reaction, *a secondary cell has to be given a charge of electricity* before it can be used — and it can be recharged after use. As it is charged, electrical energy is converted and stored as chemical energy, and then when electricity is needed the chemical energy is converted to electrical energy.

Electric circuits

Investigation 14.3 *The flow of electricity*

You will need a 4.5 volt torch battery, three torch bulbs, three bulb holders, plastic-coated stranded copper wire, paper clips, drawing pins, blocks of wood, a small screwdriver and a pair of pliers.

Proceed as follows
Make the circuit shown in Figure 14.4A. Then move the switch to connect the wires. The bulb will light as an electric current flows through it. Move the switch to break the connection. An electric current flows from the battery only if a complete path of electrical conductors connects the two terminals of the battery. The battery plus these conductors make up a circuit, which can be represented as a circuit diagram (see Figure 14.4B).

Note. In any circuit, before switching on the current, you must: (a) check that electricity cannot flow directly from one terminal of the battery to the other, along a wire that offers almost no resistance to the current and is called a short circuit, and (b) check that all connections have been made correctly (see Figure 14.5).

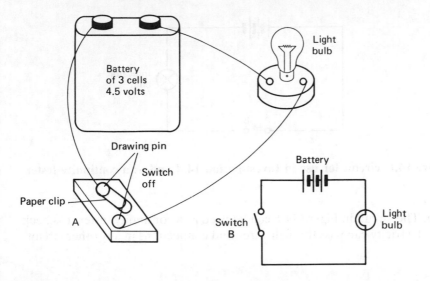

Figure 14.4 (A) an electric circuit, and (B) a circuit diagram

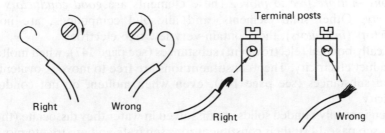

Figure 14.5 making connections in an electric circuit

Investigation 14.4 *Conductors and insulators*

You will need the same materials as for Investigation 14.3.

Proceed as follows
1 Make the circuit shown in the circuit diagram, Figure 14.6, with the wires bared of insulation for about 1 cm at A and B.
2 Touch these bared ends together and note that the bulb lights.
3 To find out what kinds of material are good conductors of electricity and what kinds are non-conductors (insulators), bridge the gap AB with different kinds of material. Keep a record of your findings.

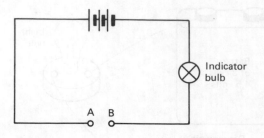

Figure 14.6 circuit for use in Investigation 14.4 and as a continuity tester

Note. The circuit in Figure 14.6 can be used as a continuity tester, to check that electricity can pass through wires and connections in any other circuit.

Conductors and non-conductors

Some solid elements (metals, and graphite — a form of carbon) contain *electrons that are free to move*. These elements are *good conductors of electricity*. Other solid elements, and all solid compounds, are non-conductors (*insulators*): they contain very few free electrons.

Ionically bonded (electrovalent) substances (see page 147), when molten do conduct electricity. Their constituent ions are free to move. Covalently bonded substances (see page 144), even when molten, do not conduct electricity.

When ionically bonded solids are dissolved in water they dissociate (they ionise, see page 143): their constituent ions separate and are free to move. The solution formed will conduct electricity (is an electrolyte). For example, sodium chloride dissociates in water:

$$NaCl \longrightarrow Na^+ + Cl^-$$

In contrast, covalently bonded solids if they dissolve in water do not ionise, so they do not conduct electricity.

Water, a covalent compound, is a very poor conductor of electricity when pure. But tap water, river water and sea water are all good conductors of electricity because they contain ions from dissolved salts.

Investigation 14.5 *Connecting bulbs in series and in parallel*

You will need the same materials as for investigation 14.3.

Proceed as follows
Try connecting three bulbs in a circuit in different ways. For each arrangement, find where the switch must be placed so that you can:

(1) switch all the bulbs on and off, and (2) switch just one of the bulbs on and off. You will find that there are two types of circuit: series circuits and parallel circuits.

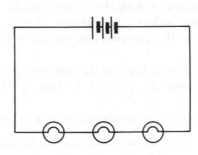

Figure 14.7 three light bulbs connected in series

When bulbs are in a chain, with each one connected to the next (as in Figure 14.7), they are said to be in series. They are in one continuous circuit: a break at any point interrupts the flow of electricity and all the lights go out. Also note that the more bulbs you arrange in series the less light is produced by each bulb.

When bulbs are connected in parallel (as in Figure 14.8), each one is as bright as if it were connected by itself and each bulb can be switched off without affecting the others. If one bulb stops working the others do not go out.

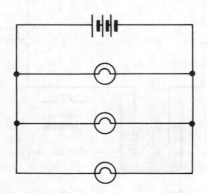

Figure 14.8 three light bulbs connected in parallel

Investigation 14.6   *Magnetic and heating effects of an electric current*

You will need the materials listed for Investigation 14.3, a small compass, a beaker and a solution of salt in water.

Proceed as follows
1 Arrange a circuit as in Figure 14.6, but with one wire parallel to a compass needle, and with the bared ends of the wires dipped in a solution of sodium chloride (see Figure 14.9).
2 What happens to the compass needle when you switch on the current?
3 The bulb lights, confirming that the salt solution conducts electricity (is an electrolyte, see page 153). Note that the bulb also gets warmer.
4 Make notes on your own observations in this investigation.

The circuit represented in Figure 14.9 can be used to demonstrate the three effects of an electric current. When an electric current flows in this circuit: (1) the bulb lights and gets warmer, (2) the compass needle moves, and (3) bubbles appear on the two wires dipped in the salt solution.

When an electric current passes through a material that offers a resistance to the flow of current, some of the electrical energy is converted to heat energy. This is a *heating effect*.

There is a magnetic field around the wire when an electric current is flowing. This *magnetic effect* is studied in more detail in Chapter 27.

When an electric current passes through a solution of an acid or a metallic salt (see page 148) a chemical change takes place in the liquid, and new chemicals are formed. This is a *chemical effect*.

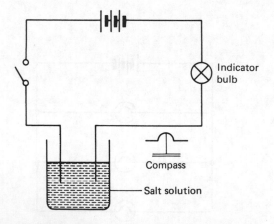

Figure 14.9 circuit used to demonstrate the heating, magnetic and chemical effects of an electric current

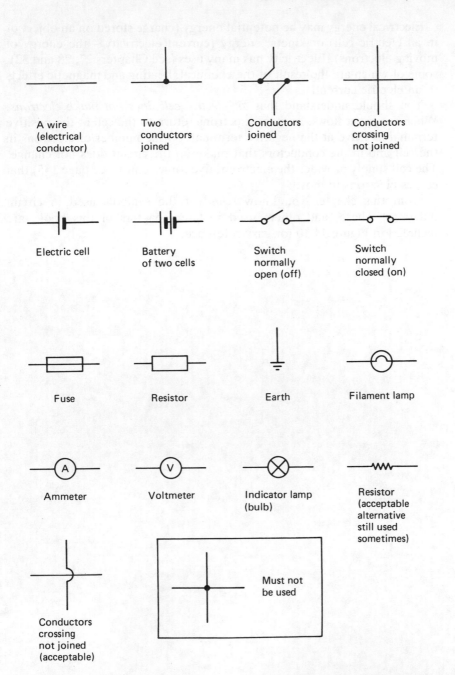

Figure 14.10 symbols used in electric circuit diagrams (including two that are acceptable but not preferred), and one that must never be used

Electrical energy may be potential energy (charge stored on an object or in an electric cell) or kinetic energy (current electricity – the energy of moving electrons). Electricity has many uses (see Chapters 27, 28 and 32), some of which are the result of the chemical, heating and magnetic effects of an electric current.

You should understand that *an electric cell does not make electrons*. When a current flows, as many electrons return to the cell at the positive terminal as leave at the negative terminal. So the number of electrons in the cell and in the conductors that make up the circuit does not change. The cell simply provides the electromotive force (e.m.f. see page 155) that causes electrons to move.

From this chapter you know some of the symbols used in circuit diagrams. These, and others used in later chapters of this book, are included in Figure 14.10 for easy reference.

⬡15 Energy from sunlight

You will remember that in sunlight green plants add oxygen to the air. In 1860, about 100 years after Priestley's work on the effects of plants and animals on the composition of air (see page 118), a German botanist Julius von Sachs demonstrated that in sunlight green leaves increase in mass.

15.1 Photosynthesis

Both the loss of oxygen from leaves and their increase in mass, in sunlight, are due to photosynthesis. This is a complicated process, not just a chemical reaction, but here is a summary of the materials used and the products:

Raw materials   *Energy*   *Products*

carbon dioxide + water $\xrightarrow[\text{in green plants}]{\text{sunlight}}$ oxygen + sugars

In many plants, the sugars produced in photosynthesis are converted to starch (a carbohydrate, see page 195) which is stored in the leaves. This is why the leaves increase in mass while photosynthesis is taking place (see Figure 15.1) — there is an increase in the amount of matter they contain.

When a plant is kept in the dark, the starch in the leaves is converted to sugars which are exported, in solution, to other parts of the plant. After a while there is little or no starch in the leaves. We say that they have been *destarched*.

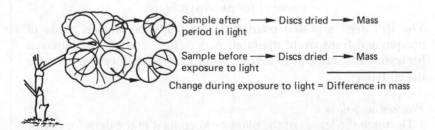

Sample after ⟶ Discs dried ⟶ Mass
period in light

Sample before ⟶ Discs dried ⟶ Mass
exposure to light

Change during exposure to light = Difference in mass

Figure 15.1 method by which Julius von Sachs, measured the increase in mass of leaf tissues exposed to sunlight. (From Barrass, R. Modern Biology Made Simple)

It is not possible, with inexpensive equipment, to demonstrate that water is used in photosynthesis. However, evidence that carbon dioxide, the green pigment chlorophyll and light are all needed is obtained in the following simple investigations. In these investigations the presence of starch in green leaves, after they have been exposed to sunlight in an atmosphere containing carbon dioxide, provides evidence that photosynthesis has taken place.

Investigation 15.1   *Testing for starch*

You will need a leaf that has been exposed to sunlight for several hours immediately before your investigation, a container in which to boil the leaf, water, a test tube, an electric kettle, ethanol (methylated spirit), a watch glass or small dish and a solution of iodine in potassium iodide. Wear eye protection.

Proceed as follows
1 Boil the leaf in water for 30 seconds to kill the tissues, stop enzyme controlled reactions (see page 207) and make the cell walls more permeable.
2 Place the leaf in ethanol in the test tube.
3 Heat the tube by standing it in a beaker of hot water from an electric kettle. Do not use a bunsen burner because the flame may ignite the alcohol vapour. The green pigment in the leaf will dissolve in the alcohol, leaving the leaf white or pale yellow.
4 Soften the leaf again in boiling water.
5 Cover the leaf with a few drops of iodine solution. Iodine stains starch blue-black. If starch is present the leaf will be stained blue-black.

Investigation 15.2   *Experiment to test the hypothesis that light is essential for photosynthesis*

You will need a potted plant with green leaves, stencils made of opaque and transparent material, paper clips, and all the materials for testing for starch as in Investigation 15.1. Remember not to use a naked flame.

Proceed as follows.
1 Destarch the leaves of the plant by keeping it in the dark for two or three days. Then keep the plant in the dark but remove one leaf and test it for starch to confirm that the leaf has been destarched.
2 Remove the plant from the dark and place an opaque stencil over both surfaces of one leaf (as in Figure 15.2A) so that light is unable to enter the covered part of the leaf.

3 Place a similar but transparent stencil over another leaf of the same plant (as in Figure 15.2B). This is your control experiment: the leaf is covered but light is not excluded.
4 Place the plant in the sunlight for several hours.
5 Label the leaves that are covered with stencils so you know which is which and then remove them from the plant and test for starch.

Note. (see Figure 15.2A$_2$ and B$_2$) that starch is present in the parts of one leaf that were not covered by the opaque stencil and in all parts of the other leaf that were covered by the transparent stencil. The observation that starch is present only in the parts of leaf A exposed to sunlight, is evidence that we can interpret in terms of the hypothesis that light energy is essential for photosynthesis. The *control experiment* enables us to check that it is the absence of light and not some other effect of the stencil that prevents photosynthesis.

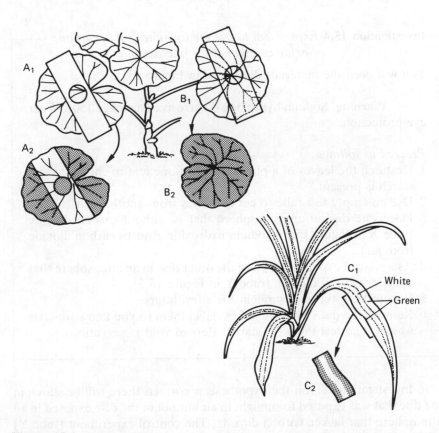

Figure 15.2 photosynthesis occurs only in parts of the leaf which (A and B) are exposed to light, and (C) contain chlorophyll. (From Barrass, R. Modern Biology Made Simple)

Investigation 15.3 *Experiment to test the hypothesis that chlorophyll is essential for photosynthesis*

You will need a plant with variegated leaves, drawing paper, a pencil, and all the materials needed for testing for starch as in Investigation 15.2. Remember not to use a naked flame. See page 164.

Proceed as follows
1 Draw a variegated leaf. Label the green and white parts.
2 Place the plant in sunlight for a few hours. Then remove the same leaf and test it for starch. Record your observations.

If in Investigation 15.3 you found starch in the green parts but not in the white parts of the leaf, your observations would be evidence in support of the hypothesis that photosynthesis is confined to the parts of the leaf containing the green pigment chlorophyll. (See Figure 15.2C.)

Investigation 15.4 *Experiment to test the hypothesis that carbon dioxide is essential for photosynthesis.*

You will need the materials illustrated in Figure 15.3.

Warning. Sodium hydroxide is corrosive (see page 150). Wear eye protection.

Proceed as follows
1 Destarch the leaves of a plant, then test one leaf to check that no starch is present.
2 Use an empty test tube to cut two discs from another leaf.
3 Place one disc in an atmosphere that is without carbon dioxide (tube X in Figure 15.3) (sodium hydroxide absorbs carbon dioxide from air).
4 As a control experiment, place the other disc in an atmosphere that contains carbon dioxide (tube Y in Figure 15.3).
5 Expose both tubes to sunlight for a few hours.
6 Remove the discs from the tubes, label them so you know which is which, then test them for starch. Record your observations.

In Investigation 15.4, if the hypothesis is correct, there will be starch in the disc that was exposed to sunlight in air but not in the disc exposed in an atmosphere that lacked carbon dioxide. The control experiment (tube Y) enables you to check that it is the absence of carbon dioxide, not some other factor related to the design of the experiment, that prevents photosynthesis from taking place in tube X. Did your investigation provide

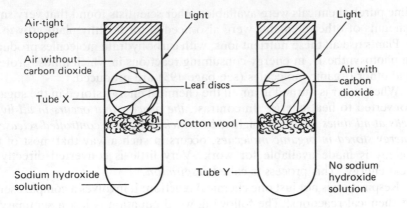

**Figure 15.3 photosynthesis occurs only if the air contains carbon dioxide.
(From Barrass, R.** Modern Biology Made Simple)

evidence which you could interpret in terms of the hypothesis that carbon
dioxide is essential for photosynthesis?

Every experiment should be planned so there is an appropriate control
experiment, and both experiments should be conducted at the same time if
possible.

The results from the last three investigations may provide evidence
which causes you to think that the hypotheses tested are probably correct,
but the evidence does not prove that these hypotheses are correct. There
may be other explanations for the results obtained. This is true of most
experiments. *The results of one experiment may provide evidence but they
do not provide proof.*

15.2 Matter and energy used by all life on Earth

Green plants use energy and matter

Green flowering plants (see Figure 5.5), and other organisms that have the
green pigment chlorophyll (for example see Figure 3.2), can capture and
use energy from sunlight. In photosynthesis, light energy is captured and
stored as chemical energy in carbohydrate molecules. This stored energy
may be used later to do work. For example, energy is used when the roots
of a flowering plant take in nutrients from the soil.

In 1860, two German scientists, Sachs and Knop, tried to find out which
elements absorbed by plant roots were essential for normal plant growth.
They grew plants, without soil, in solutions containing different combinations of soluble salts. The plants did not grow properly unless the solutions
contained calcium, potassium, magnesium, nitrate, phosphate and sulphate ions, and also very small amounts of iron. In the 1920s, by which

time purer chemicals were available, other scientists found that very small amounts of other elements were also needed for healthy plant growth.

Plants use all these nutrient ions, with carbohydrate molecules produced in photosynthesis, in energy-consuming reactions in which lipids, proteins and other organic molecules (see page 195) are produced.

When sugar is burned in air all the chemical energy stored in the sugar is converted to heat energy. In contrast, *the process that occurs in all living cells at all times* (day and night), *by which there is a controlled release of energy stored in organic molecues*, occurs in such a way that most of the energy is made available for work. Very little is converted directly to heat energy. This process is called *respiration*.

Respiration is not just one chemical reaction: it involves a complex series of chemical reactions. The following word equation is just a summary of the materials used and the products.

Raw materials	in all living cells	*Products*
sugar + oxygen	$\xrightarrow[\text{day and night}]{\text{at all times,}}$	carbon + water + energy dioxide

Respiration and photosynthesis in the cells of a leaf

In Figure 15.4 note the parts of a leaf: (A) the stalk and blade, (B) the leaf blade cut across, and (C and D) the different kinds of cells visible in a cross section.

The outermost cells of a leaf form a skin, one cell thick, called the epidermis. Note that the cells of the upper epidermis are closely packed but that between some cells of the lower epidermis there are gaps called *stomata*. Through these stomata the outside air is continuous with air that fills spaces between cells inside the leaf.

Most cells inside the leaf, called parenchyma cells, contain chloroplasts. But note that the leaf also contains tubes (formed from cells). In some of these water and mineral ions are imported, via the stem, from the roots. In others organic molecules produced in the leaf are exported to other parts of the plant.

All leaf cells, like all other living cells, respire at all times. Photosynthesis, however, occurs only in those cells that have chloroplasts, and only in the presence of light. In the daylight photosynthesis usually proceeds much faster than respiration. More carbon dioxide is used in photosynthesis than is produced in respiration; and more oxygen is produced in photosynthesis than is used in respiration. This is why carbon dioxide diffuses into the leaf and oxygen diffuses out (see Figure 15.4C). Note also that, except in an atmosphere saturated with water vapour, water vapour diffuses out of the leaf.

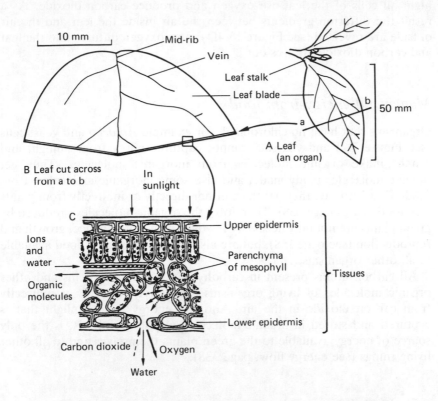

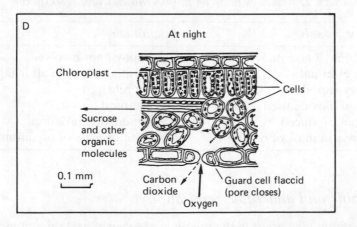

Figure 15.4 different kinds of cell in the leaf of a flowering plant. And gas exchange in sunlight (C) and at night (D). (From Barrass, R. Modern Biology Made Simple)

Although photosynthesis stops at sundown, respiration continues. At night all cells of the leaf use oxygen and produce carbon dioxide. As a result the diffusion gradients between the air inside the leaf and the air outside are reversed (see Figure 15.4D): now oxygen diffuses into the leaf and carbon dioxide diffuses out.

Matter and energy from food

Organisms that have no chlorophyll, for example *Amoeba* and yeast cells (see Figure 7.2) and yourself, cannot capture energy from sunlight and cannot make organic molecules from inorganic molecules. They get organic molecules ready made, and also some inorganic molecules, from food. That is to say, they get them either directly or indirectly from plants (see food chains page 338). Therefore, the organic molecules produced by green plants are not only necessary for their own maintenance, growth and reproduction (see page 168), but are also the only source of food available to all other organisms.

All carbon atoms present in carbohydrates, lipids, proteins and other organic molecules of living organisms are derived directly or indirectly from carbon dioxide in the air. And the energy from sunlight that is captured and stored in organic molecules in photosynthesis, is the only source of energy available to the green plants themselves and to all other living things (see energy flow, page 338).

Table 15.1   *Differences between photosynthesis and respiration*

Photosynthesis	Respiration
Chlorophyll essential	Chlorophyll not involved
Light essential	Occurs at all times in all living cells
Energy stored	Energy released
Carbon dioxide used	Oxygen used
Oxygen produced	Carbon dioxide produced
Increase in mass of organism	Decrease in mass of organism

Aerobic and anaerobic respiration

Respiration may occur in the presence of oxygen, as it does at most times in all cells of your body. However, when you are very active, some cells are using a lot of energy and may not get enough oxygen for aerobic respiration. Then respiration may continue in a shortage or even an absence of oxygen. Such respiration without oxygen is called anaerobic respiration.

Some animals live in places where there is very little oxygen (as do tapeworms living in the small intestine of their host, see page 349). Such animals obtain almost all their energy from anaerobic respiration. Some fungi, including yeast (see page 73) respire aerobically when oxygen is present but can respire anaerobically if there is no oxygen.

In both aerobic and anaerobic respiration, carbon dioxide is produced and there is a loss of heat. In experiments, therefore, carbon dioxide production and loss of heat are expected if respiration is taking place.

Investigation 15.5 *Anaerobic respiration of yeast*

You will need glucose solution (5 per cent glucose by volume), dried bakers' or brewers' yeast, and the other materials illustrated in Figure 15.5.

Proceed as follows

1 Boil the glucose to remove any dissolved oxygen. Cool, then use to make a suspension of yeast (10 per cent yeast by volume).

2 Arrange the apparatus as in Figure 15.5. The liquid paraffin in tubes A and B is to exclude air.

3 Keep the tubes at about 35 °C and note that the lime water becomes cloudy (see page 115) as gas from tube A bubbles through it, indicating that carbon dioxide is produced in tube A even in the absence of oxygen.

4 As a control investigation, so that you can be sure that it is the yeast cells that are producing the carbon dioxide, follow the above procedure but use glucose solution without yeast.

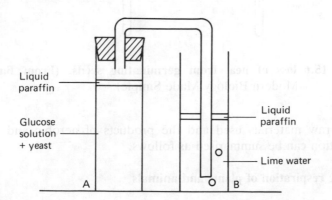

Figure 15.5 evolution of carbon dioxide by yeast in the absence of oxygen.
(From Barrass, R. Modern Biology Made Simple)

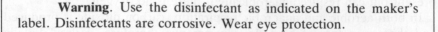

Investigation 15.6   *Loss of heat by germinating seeds*

You will need three vacuum flasks, three thermometers, disinfectant, cotton wool and pea seeds prepared as shown in Figure 15.6.

Warning. Use the disinfectant as indicated on the maker's label. Disinfectants are corrosive. Wear eye protection.

Proceed as follows
Arrange the flasks as in Figure 15.6.
Flasks B and C are your controls. Microbes on the surface of the peas are killed by disinfectant and peas are killed by boiling. If the temperature rises in flask A more than in B, and there is no rise in C, you can conclude that the loss of heat from the living seeds and living microbes is an indication of chemical activity, as you would expect if respiration occurs in all living cells at all times.

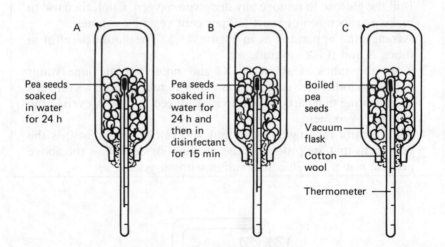

Figure 15.6 loss of heat from germinating seeds. (From Barrass, R. Modern Biology Made Simple)

The raw materials used and the products of aerobic and anaerobic respiration can be summarised as follows.

Aerobic respiration of plants and animals

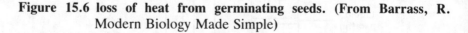

$$C_6H_{12}O_6 \ + \ O_2 \longrightarrow CO_2 \ + \ 6H_2O \ + \ \text{Energy}$$

glucose   carbon   water
                               dioxide

Anaerobic respiration (alcoholic fermentation) in yeast

$$C_6H_{12}O_6 \longrightarrow 2C_2H_5OH + 2CO_2 + Energy$$

glucose ethanol carbon dioxide

Carbon dioxide from bakers' yeast causes bread to rise; then the heat in baking kills the yeast. In brewing, fermentation stops when all the sugar has been used or when so much alcohol has been produced that it kills the yeast. The bubbles of carbon dioxide gas coming out of solution when the pressure in a bottle is reduced makes some alcoholic drinks fizzy.

Anaerobic respiration in animals

$$C_6H_{12}O_6 \longrightarrow 2C_3H_6O_3 + Energy$$

glucose lactic acid

Some animals excrete lactic acid. In others, including people, any lactic acid produced in muscles during exercise is oxidised when oxygen is available — after the exercise has stopped. That is to say, during exercise an oxygen debt is incurred — and afterwards it is repaid.

Much more energy is obtained from aerobic than from anaerobic respiration. The summary equations may cause you to think that respiration is just one chemical reaction but it is not. There is a complex series of reactions which makes possible the controlled release of energy, a little at a time. The first reactions are identical in aerobic and anaerobic respiration. Glucose, a six-carbon molecule, is broken into three-carbon molecules and some energy is released. But in anaerobic respiration most of this energy is used in the production of either ethanol (as in yeast) or lactic acid (as in your own muscles during exercise). In contrast, in aerobic respiration all the energy released in these first reactions is available to do other work in the cell. Then even more energy is released in further reactions in which each of the three-carbon molecules is completely oxidised to carbon dioxide and water.

16 Sound and hearing

16.1 How sound travels

Every sound is caused by something vibrating. For example, when you beat a drum its skin vibrates. When a guitar is played its strings vibrate. These vibrations cause the surrounding air to vibrate, and vibrations in the air cause your eardrums to vibrate — and you hear sounds.

Sound vibrations can pass through air because air is elastic. (When we say something is elastic, we mean that if the material is compressed it returns to its former shape when the pressure is removed.) When the molecules in air are squeezed together (when air is compressed), they move apart again (the air expands) as soon as the pressure is removed.

Sound vibrations can travel through any elastic material, and most materials are elastic to some extent. Therefore, sound travels through most materials. If there is no material to vibrate, sound cannot travel — sound cannot travel through a vacuum.

It is important to understand that although vibrations move through the air when sound travels, the air itself does not move from the sound source to the ear. You cannot see air, so to help you visualise the movement of sound through air it is useful to compare it with the movement of something you can see — ripples on a water surface.

Investigation 16.1 *Watching water waves*

You will need a large tray, water, tissue paper and a glass rod.

Proceed as follows
1 Partly fill the tray with water.
2 Let one drop of water fall from the glass rod on to the middle of the still water surface.
3 Watch the movement of the ripples produced after the drop of water hits the water surface. Note that these ripples move out in circles from the place where the water was disturbed — their source.
4 Now place some small pieces of tissue paper on the water surface. Let another drop of water fall onto the middle of the still water surface. Note how the ripples affect the pieces of paper. They bob up and down, as ripples pass, but they do not move outwards. This shows that the water is not moved horizontally by the ripples.

Wave motion

The following terms are used in describing all wave motion:

Amplitude: *the maximum displacement of a wave from the undisturbed level* (as in Figure 16.1)

Wavelength: *the distance travelled in one vibration of a wave.* This is the distance between two similar points on a wave, and is indicated in Figure 16.1 as the distance between successive crests.

Frequency: *the number of times a wave vibrates in one second.* One vibration (one complete wave) is called one cycle, and frequency is measured in cycles per second. In SI units (see page 55) one cycle per second is called one hertz (symbol Hz). The frequency of the wave represented in Figure 16.1 is 10 cycles per second (10 Hz).

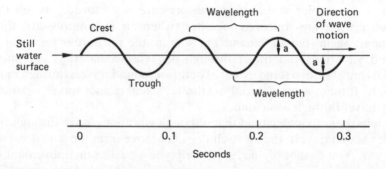

Figure 16.1 a water surface, undisturbed (straight line) and disturbed by a wave: a = amplitude; frequency 10 Hz (ten waves per second)

The *velocity* of a wave is the distance it travels divided by the time taken (see page 59). The distance travelled by a wave each second is calculated by multiplying the wavelength by the number of wavelengths per second.

velocity of a wave = wavelength × frequency

Ripples on a water surface are called water waves. You have seen that water waves pass without the water moving horizontally. Obviously, sound waves are different from water waves but they also pass through matter (for example, through air or water) without the molecules of that matter moving away from the source of the sound.

All wave motion involves the transfer of energy, not matter. An input of energy is needed to start a wave, and the wave transfers energy away from the source. The greater the amplitude of a mechanical wave, the more energy it carries. As the distance from the source increases, the wavelength and frequency do not change but as energy is dissipated the amplitude decreases.

Floating objects are moved up and down by ripples on a water surface – at right angles to the direction of the ripples. Waves of this kind are called *transverse waves*.

A ripple tank is a shallow clear-glass tray used to study waves. A straight edge is moved up and down, striking the water surface at regular intervals to produce ripples. Shadows of the ripples are observed on a white screen below the tank, and some properties of wave motion can be demonstrated. All wave motion can be reflected, refracted and diffracted, and is subject to interference (see Figure 16.2). For example, an echo results from the reflection of sound waves.

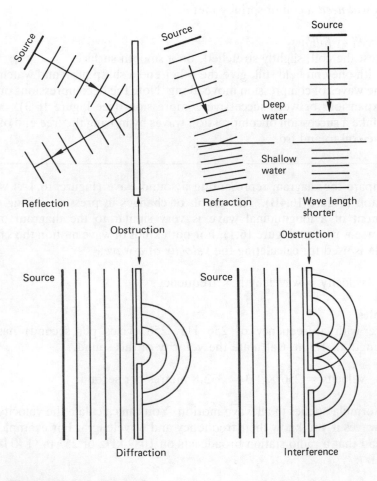

Figure 16.2 shadows of water waves on a ripple tank

Sound waves

Sound travels faster through solids and liquids, in which the molecules are close together, than through gases, in which the molecules are further apart, because when sound travels through a material the molecules of the material alternately move together and apart.

If the molecules were visible you would see them moving to and fro, away from and then towards the sound source. This is in contrast to transverse wave motion (see page 177) in which the molecules move at right angles to the wave motion. The movement of sound vibrations through a material is a kind of wave motion called *longitudinal wave motion*.

Investigation 16.2   *Observing longitudinal wave motion*

You will need a coil of springy metal.

Proceed as follows
1 Rest the coil, slightly stretched, on a smooth surface.
2 With one end held still, give the other end a sharp push and watch the waves of compression move along. Note the decompressions or expansions between successive compressions (see Figure 16.3).
3 Make a succession of compression waves by moving the free end of the coil to and fro.

Compare the diagram representing a sound wave (Figure 16.4A) with the graph (Figure 16.4B). This graph of changes in pressure during the movement of a longitudinal wave is very similar to the diagram of a transverse wave (see Figure 16.1). For both kinds of wave motion the same formula is used for calculating the *velocity of a wave*:

$$\text{velocity} = \text{wavelength} \times \text{frequency}$$

Examples
A note with a frequency of 256 Hz (256 cycles per second) has a wavelength of 1.3 m. Calculate the velocity of this sound.

$$\text{velocity} = 256 \times 1.3 = 332.8 \text{ metres per second}$$

This formula applies to all wave motion. You can calculate the velocity of radio waves if you know their frequency and wavelength. For example, if you read that a radio station broadcasts on 1089 kHz or 257 m (1 kHz = 1000 Hz):

$$\begin{aligned} \text{velocity} &= 1089 \times 1000 \times 257 \\ &= 2.8 \times 10^8 \text{ metres per second} \end{aligned}$$

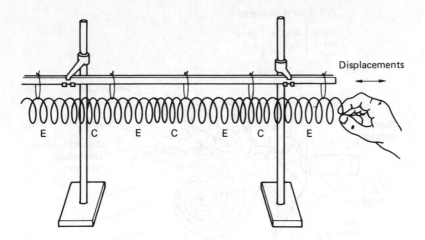

Figure 16.3 **compression waves in a loosely wound coil of wire, caused by displacements of one end of the coil. Note the alternation of compressions (C) and decompressions or expansions (E) along the coil, which in this drawing is suppported at intervals by strings, but which in your investigation rests on a smooth surface. (Based on Keighley et al** Mastering Physics)

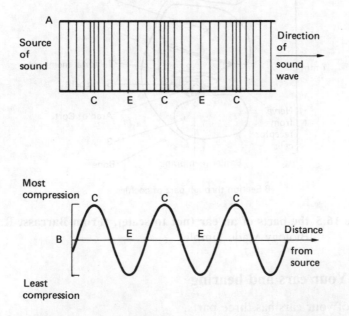

Figure 16.4 **longitudinal wave motion: (A) diagram representing a sound wave (C) = compression; (E) = expansion (decompression or rarefaction); (B) a graph of changes in pressure during transmission of this wave**

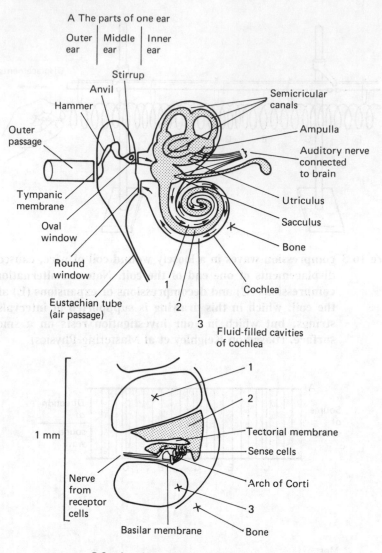

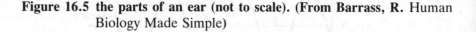

B Section through part of cochlea

Figure 16.5 the parts of an ear (not to scale). (From Barrass, R. Human Biology Made Simple)

16.2 Your ears and hearing

Each of your ears has three parts.

1 The outer ear includes the part you see on the outside (not shown in Figure 16.5) and the passage that admits sound waves to the eardrum (see Figure 16.5).

2 The middle ear is connected to your pharynx (at the back of your mouth cavity) by an air-filled tube, the eustachian tube (see Figure 16.5). As a result, your middle ear contains air at the same pressure as the outside air. The three bones (the hammer, anvil and stirrup, see Figure 16.5) move as the eardrum vibrates and carry vibrations to the membrane of the oval window, which also vibrates.
3 The inner ear comprises: (a) *an organ of balance* (the utriculus, sacculus and *semicircular canals* with their ampullae, see Figure 16.5), and (b) the *cochlea* which, with the outer and middle ear, is an *organ of hearing*.

The cochlea is a spiral tube full of fluid. Because this fluid is incompressible (see page 67), vibrations of the membrane stretched across the oval window are transmitted (as indicated by arrows in the diagram) to the membrane stretched across the round window. Vibrations caused by sounds of different pitch stimulate sense cells at different distances along the cochlea. Low notes stimulate sense cells near the tip of the cochlea, high notes stimulate cells towards the base, and notes of intermediate frequencies those in between. The sense cells stimulated produce impulses in the nerve to the brain, and the brain interprets these signals.

It is easier to locate the source of a sound with two ears than it would be with one, because you are able to compare the loudness of the sounds at the two sides of your head.

Music and noise

In terms of wave motion, music is made up of regular vibrations. The pitch of a note (whether it is high or low) depends on the frequency of the vibration. Low-frequency (slow) vibrations produce low notes and high-frequency (fast) vibrations produce high notes. The audible range of people is from about 20 Hz to about 20 kHz but a higher pitched whistle can be heard by dogs — and bats emit and detect ultrasonic waves which they use in echo-location.

The larger the amplitude of the vibrations the louder the note.

In general, a sound is called a noise if it is not wanted or if it is too loud. People are becoming increasingly concerned about the harmful effects of noise. Health can be adversely affected if noises stop people from sleeping. And very loud or continuous noises can cause permanent damage to hearing. Anyone shooting a rifle or using a pneumatic drill, for example, should wear ear plugs to protect their ears.

Sound is a form of mechanical energy, and the rate of change of one form of energy to another can be measured in watts (see page 113). For example, loudspeakers convert electrical energy to sound energy and are rated in watts. However, the watt is not a suitable unit for measuring the effects of noise on people. One reason for this is that our ears are sensitive to sounds over a wide range. The loudest sound we can listen to without feeling pain is about a million million times the power of the quietest sound we can just hear. Another reason is that our ears are less sensitive to increases in sound as the sound becomes louder. For example, if the power of a sound is increased ten times it does not seem ten times as loud to us.

For these reasons, a special scale of loudness based on a unit called the decibel has been devised so that the loudness of different noises can be measured in relation to the hearing ability of people. The faintest sound that can be heard by a person with normal hearing is called the *threshold of hearing*. This is taken as zero (log 1) on the decibel scale. Ten decibels is ten times this threshold level, 20 decibels 100 times, 30 decibels 1000 times, and so on. Such a scale is called a logarithmic scale.

On the decibel scale a whisper is 30, talking 60, the noise of heavy traffic 80, a pneumatic road drill 95, and an aeroplane at take off 110.

The decibel scale is accepted internationally for measuring the possible effects of noise on people. For example, in the UK laws have been passed that state the maximum permissible noise levels (in decibels) of vehicles.

Although widely used, the decibel scale is not ideal for measuring the effects of noise because some kinds of noise are less acceptable than others. For example, high-pitched sounds at quite a low noise level — such as may be caused by scraping a fingernail on a hard surface — can be particularly annoying.

Attempts to minimise the harmful effects of noise on people are being made in most industrialised countries. Where possible, noise levels should be reduced. When this is impossible, people can be protected by sound insulation, or by using ear plugs or ear muffs.

(17) Light and seeing

17.1 Light travels in straight lines

Light passes through *transparent* material (for example, glass) but not through *opaque* material (for example, wood). If an opaque object is placed in front of a point light source, in a darkened room, a sharp edged shadow is formed in the region of less intense light — indicating that *light travels in straight lines* (see Figure 17.1). You can imagine light spreading from the point source in all directions, as rays, in straight lines. However, in a diagram (such as Figure 17.1) only those rays needed to show the position of the shadow are represented.

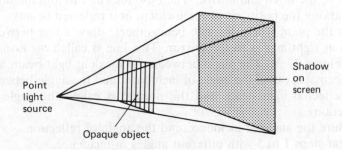

Figure 17.1 light travels in straight lines

Reflection

When a light shines on a surface some light may be absorbed. The rest bounces off. The light striking a surface is called *incident light*, and light bouncing off is *reflected light*. From a rough surface light is scattered, but from a smooth surface light is reflected in a regular manner and we call such an object a mirror (see Figure 17.2).

Investigation 17.1   *The two laws of reflection*

You will need a flat mirror, a torch with its glass largely covered so it gives only a narrow beam of light, paper, a sharp pencil, a ruler and a protractor.

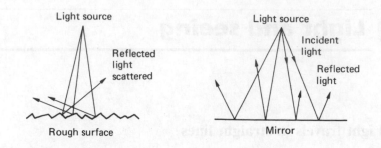

Figure 17.2 reflection

Proceed as follows

1 Place the mirror upright on the paper and draw a line using the back of the mirror as a straight edge. The back of the mirror is the reflecting surface.

2 Shine the torch at the mirror so that you can observe the incident and reflected beams. Mark their positions with dots.

3 Remove the torch and mirror. Then use the ruler to join the dots, so marking the positions of the incident and reflected beams.

4 From the point where the two beams meet, draw a line between them at right angles to the mirror. This line is called *the normal* (see Figure 17.3). The angle between the incident light beam and the normal is called the angle of incidence, and the angle between the reflected light beam and the normal is called the angle of reflection.

5 Measure the angle of incidence and the angle of reflection.

6 Repeat steps 1 to 5 with different angles of incidence.

Each time you carry out this investigation you will find that, because light travels in straight lines, *the angle of incidence is equal to the angle of reflection*. This is the *first law of reflection*.

You were able to draw the incident beam, the reflected beam and the normal on a sheet of paper because they all lay on the same flat surface of plane. This is the *second law or reflection: the incident ray, the reflected ray and the normal all lie in the same plane*.

Refraction

When a ray of light travelling through air, a transparent material, strikes the surface of another transparent material at an angle, the ray of light changes direction. This bending of light is called *refraction*, and the ray of light is said to be *refracted*.

When light travels from a less dense to a more dense material, for example from air to water or glass, it is bent towards the normal (see

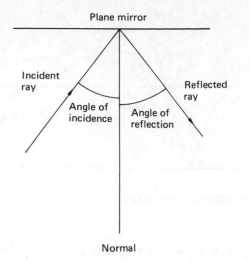

Figure 17.3 **the angle of incidence is equal to the angle of reflection**

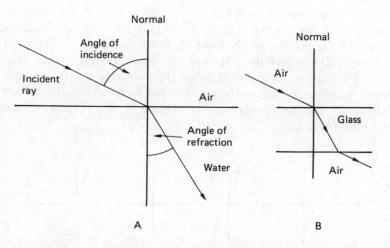

Figure 17.4 **refraction of light travelling from air to a more dense material (water or glass, A and B), and from glass to a less dense material (air, B)**

Figure 17.4A). When light travels from a more dense to a less dense material it is bent away from the normal (see Figure 17.4B).

One effect of refraction is seen when we look at objects that are under water. A straight rod that is partly immersed in water appears to be bent (see Figure 17.5). When your eye receives light rays from the end of the rod (marked A in Figure 17.5), they seem to come from point B.

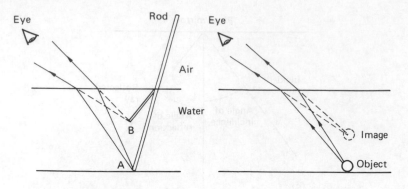

Figure 17.5 two effects of refraction

Another effect of refraction is that water in a swimming pool looks shallower than it really is (see Figure 17.5).

Total internal reflection

When a ray of light travels from a more dense into a less dense material it is bent away from the normal (Figure 17.6A). If the angle of incidence is increased the ray is bent further from the normal (Figure 17.6B), until an angle is reached at which the refracted ray just grazes the surface (Figure 17.6C). This angle of incidence is called the *critical angle*. If the angle of incidence is further increased, the ray does not pass through the surface — it is reflected within the denser material (Figure 17.6D). This is called *total internal reflection*.

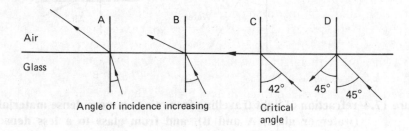

Figure 17.6 refraction of light (A to C) and total internal reflection (D) above the critical angle (C)

One application of total internal reflection is in the use of two triangular glass prisms to make a periscope (see Figure 17.7). Light strikes the first face of the upper prism at right angles, so there is no refraction at this face. The light then strikes the oblique face at an angle of 45°. This is greater

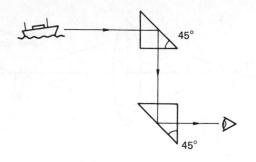

Figure 17.7 a periscope

than the critical angle, which for glass to air is about 42°, so the light is totally internally reflected. As the angle of incidence is 45°, the angle of reflection will also be 45°, so the ray will be turned through 90° and will strike the lower face of the first prism at 90° and will not be refracted at this face. The ray travels in a similar way through the lower prism and then enters the eye of the observer.

Lenses

Our knowledge of the bending of light as it travels from one transparent material to another is also used in making lenses. Many instruments, including cameras, projectors and spectacles, contain lenses.

Lenses are made in many different shapes but there are two main types. A lens that is thicker in the middle than at the edge is called a *convex lens* or, because it converges light to a focus (see Figure 17.8A), a converging lens. A lens that is thinner in the middle than at the edge is called a *concave lens*. Because this diverges rays so that they appear to come from a point (see figure 17.8B), such a lens is called a diverging lens.

We can assume that light from the sun, because it is such a distant object, reaches Earth as parallel rays. The point at which parallel rays of light are focused by a lens is called the principal focus of the lens; and the distance from the lens to the principal focus is called the *focal length* of the lens (see Figure 17.9).

You can easily find the focal length of a convex lens by focusing an image of a brightly illuminated window on to a wall or card at the opposite end of the room. The distance from the image to the lens is the focal length of the lens. The shorter the focal length of a lens the more powerful it it said to be. But be careful – fires can be caused this way.

The images produced by convex lenses. In some instruments, including a camera, a lens is used to focus an image of an object on to a screen. Such an image is called a *real image*.

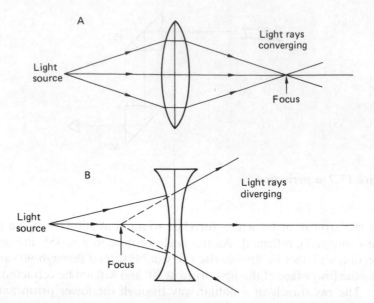

Figure 17.8 two kinds of lens: (A) biconvex, a converging lens; and (B) biconcave, a diverging lens

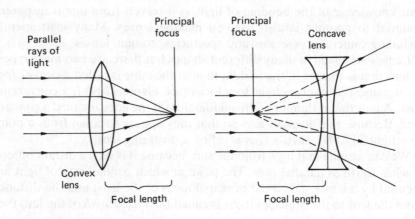

Figure 17.9 focal length

You can determine the position of an image produced by a particular lens by drawing a ray diagram to scale. To do this you must trace the path of two rays coming from a point on the object:

1 The ray that is parallel to the axis of the lens (see Figure 17.10) which will pass through the lens and then through the principal focus.

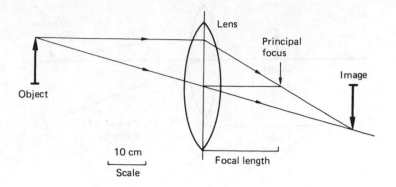

Figure 17.10 a real, inverted image

2 The ray that passes through the centre of the lens and continues without being refracted.

Example. Where will the image be formed when an object is placed 45 cm in front of a convex lens that has a focal length of 20 cm (see Figure 17.10). **Note** that a convex lens produces a real, inverted, image.

A convex lens can be used as a magnifying glass. If you move an object towards such a lens, a point is reached at which it is not possible to form a real image. This point is at a distance of one focal length from the lens. If you draw a ray diagram with the object within this distance you will see that the rays diverge after passing through the lens (see Figure 17.11A). And if you look into the lens, at an object that is within the focal length, you will see a magnified image of the object. You can determine the position and size of the image if you trace the rays back through the lens (as in Figure 17.11B). Note that this is not a real image. It is magnified and the right way up, and is called a *virtual image*.

Optical instruments

The study of light and sight is called optics. And instruments such as cameras, microscopes, spectacles and telescopes, which allow us to use our knowledge of the properties of light, are called optical instruments.

A pinhole camera can be made from a cardboard box, tinfoil and tissue paper (see Figure 17.12A). But the cameras used in photography have lenses that focus light on to light-sensitive film (see Figure 17.12B).

In all except the simplest cameras the lens can be moved so that light from objects at different distances from the camera can be focused on the film. The size of the opening at the front of the camera (the aperture) can be adjusted to compensate for changes in the brightness of the object. For example, on a dull day the aperture can be made larger so that enough light

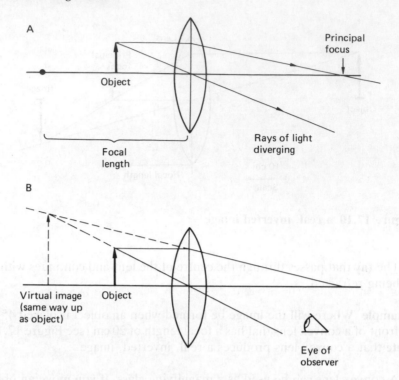

Figure 17.11 object within the focal length of a lens: (A) rays of light from object diverging after passing through lens; and (B) virtual image (magnified) when lens used as a magnifying glass

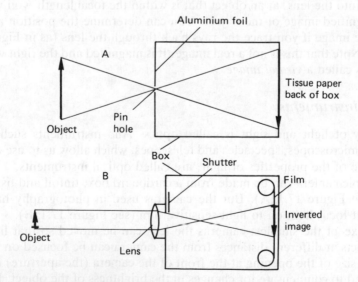

Figure 17.12 (A) pinhole camera; and (B) parts of a box camera

enters the camera for a photograph to be taken. The shutter, which allows light through the aperture for a definite time, can also be adjusted. For most photographs it is open for only a fraction of a second.

Lenses are also used in magnifying glasses (see page 71) and in spectacles (see page 193). Prisms are used not only in periscopes (see page 187) but also in binoculars and in some cameras.

It is also possible to pass light along flexible transparent tubes, in which light is internally reflected (see Figure 17.6D). Such tubes, consisting of bundles of very fine fibres, have many uses (in what is called *fibre optics*). For example, a doctor can insert a fibre optic tube by way of the mouth and oesophagus (see figure 18.5) so that the inner surface of the stomach can be examined. Another use is in telecommunications where pulses of light can be sent along fibres, instead of sending pulses of electricity along wires.

17.2 Your eyes and seeing

Light enters the eye through the conjunctiva (the transparent skin) and cornea (the transparent part of the fibrous coat) and is focused by the lens on to the retina (see Figure 17.13).

There are two kinds of light–sensitive cells in the retina: rods which are not colour sensitive, and cones which are colour sensitive but do not function at low light intensities. This is why you cannot see colours at night. The cones are most abundant, and the rods least abundant, on or near the optical axis (the dotted line in Figure 17.13). This is why you see images

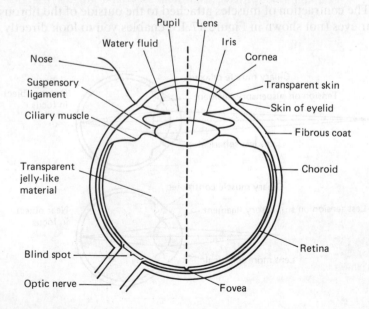

Figure 17.13 horizontal section through right eye (not to scale). (From Barrass, R. Human Biology Made Simple)

and colours most clearly in bright light when light is sharply focused on the centre of the retina (see Figure 17.14).

The choroid layer, immediately below the retina, contains pigmented cells which absorb the light that passes through the retina. Like the black surfaces inside a camera, this prevents the scattering of light that would blur the image. The choroid also contains the eye's blood vessels.

Nerve cells in the retina and optic nerve carry signals from the light–sensitive cells of the retina to the brain. Where the nerve cells converge on the optic nerve there are no sense cells and this is why you have a blind spot in each eye. Close your left eye and look at the cross below. Move the book towards you, slowly, and note that the black dot disappears and then reappears. The image of the black dot is not visible when this is formed on the blind spot.

+   ●

As in a camera, the lens in the eye is convex. But focusing involves not a change in the position of the lens (as is essential with the rigid lens of a camera) but a change in its shape. The lens of your eye is elastic and is flattest when distant objects are in focus (Figure 17.14A). Contraction of the ciliary muscles reduces the tension on the suspensory ligament and, as a result of its elasticity, the lens becomes closer to being spherical when nearby objects are in focus (Figure 17.14B). There is less eyestrain, therefore, if you do not hold the page too near your eyes when reading.

You see most clearly the objects which are immediately in front of your eyes. The contraction of muscles attached to the outside of the fibrous coat of your eyes (not shown in Figure 17.13) enables you to look directly at an

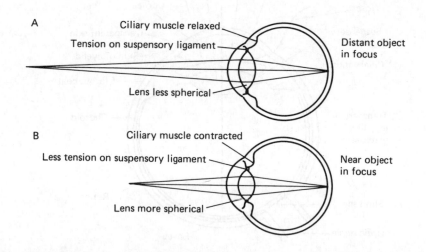

Figure 17.14 eye focused: (A) on a distant object, and (B) on a nearby object. (From Barrass, R. Modern Biology Made Simple)

object with both eyes at the same time, and to keep your eyes on a moving object when your head is still or upon an object that is still when your head is moving.

Control over the amount of light entering each eye depends partly on behaviour, as when you turn away from a bright light or shade your eyes. It also depends on changes in the size of the opening, the pupil, due to contraction of the iris muscles. As with a camera, the sharpest image is formed in bright light with a small aperture.

People who cannot see distant objects clearly, because the light from them is focused in front of the retina, are said to be shortsighted. This defect of vision can be corrected by wearing spectacles with appropriate concave lenses. People who cannot see nearby objects clearly, because light from them is focused behind the retina, are said to be longsighted. This defect of vision can be corrected by wearing spectacles with appropriate convex lenses.

18 Matter and energy from your food

18.1 The matter of life

Of the 94 elements that occur naturally, only about 22 (all of low atomic mass) are essential constituents of living organisms.

The most abundant constituent of any organism is water. Other chemical compounds are present in solution in the water: for example, sodium chloride and carbon dioxide. Oxygen molecules are also present in solution. These four substances, like many others that occur both inside and outside living organisms, are called inorganic molecules.

Organisms also contain organic molecules, so called because they are produced in nature, from inorganic molecules, only in the bodies of living organisms. Organic molecules are compounds of carbon, hydrogen, and oxygen; and some also contain other elements. Many organic molecules contain a very large number of atoms, and so have a very large mass. These macromolecules are part of the complex organisation maintained throughout the life of each organism.

Carbohydrates are compounds of carbon, hydrogen and oxygen in which the ratio of hydrogen to oxygen atoms is 2:1 (as in water). For example, glucose (formula $C_6H_{12}O_6$) is a 6-carbon or hexose sugar. Its atoms are arranged in one six-sided ring (see Figure 18.1), so it is called a monosaccharide (mono = one). Maltose, another sugar, is formed from two glucose molecules. It is a disaccharide (di = two). This type of chemical reaction, in which two molecules combine and water is formed (see Figure 18.2) is called a condensation reaction.

Starch (a food reserve in plants), cellulose (the main constituent of plant cell walls), and glycogen (a food reserve in your muscles and liver) are all

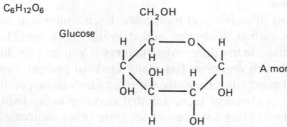

Figure 18.1 glucose, a six-sided ring

Two molecules of glucose

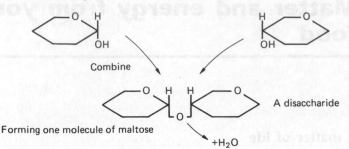

Combine

A disaccharide

Forming one molecule of maltose

+H₂O

Amylose (starch)

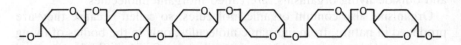

and cellulose are polysaccharides (with many glucose units)

Figure 18.2 larger carbohydrate molecules formed by condensation of small molecules of sugars, such as glucose, with the elimination of a molecule of water as each new link is added to the chain. (From Barrass, R. Human Biology Made Simple)

natural polymers (see page 294). Each one consists of 6-carbon rings, joined by condensation, like links in a chain (see Figure 18.2). They are called polysaccharides (poly = many).

Lipids (fats and oils) are also compounds of carbon, hydrogen and oxygen, but the hydrogen and oxygen are not in the same proportions as in water. Each lipid molecule is formed from one molecule of glycerol linked to three fatty acid molecules, so it is called a triglyceride (tri = 3). Different lipids contain different combinations of fatty acids. Some lipids are constituents of cell membranes. Others are accumulated in particular cells as food reserves.

Proteins are chains of amino acid molecules. Each amino acid contains nitrogen as well as carbon, hydrogen and oxygen. There are 23 amino acids. You could make many different structures if you had 23 different kinds of brick; and each organism has many kinds of protein molecule. Some are structural materials; others are enzymes which make possible the complex sequences of chemical reactions that occur in every living cell. These enzymes help to bring together or separate other molecules.

The study of the chemical substances and chemical reactions that occur in living cells, is called *biochemistry*. Note, however, that all matter and energy in living organisms conform to the same laws of chemistry and physics as the matter and energy outside them. A knowledge of physics and chemistry will therefore contribute to your understanding of biology.

18.2 Your food needs

The materials and energy needed by all the cells of your body are obtained from the variety of foods you eat. Good feeding is the basis for good health, so it is worth knowing what kinds of food you need, how much food you need each day, how people differ in their food needs, and which foods can be harmful. *The study of the food needs of organisms* is called *nutrition*.

Body-building foods

Proteins. Each kind of protein in your body contains some, not all, of 20 different amino acids. Some of these amino acids can be formed in the body (for example, by converting one kind of amino acid to another). Others, because they must be present in your diet, are called essential amino acids.

Because most plant proteins contain only small amounts of some essential amino acids they are called second-class proteins. First-class proteins, which are good sources of all the essential amino acids, are obtained from eggs, milk and cheese, lean meat, and fish.

People who have enough to eat and yet do not get enough of one or more of the essential amino acids, suffer from *protein deficiency*. This disease is known by a Ghanaian word *kwashiorkor*, which means 'the sickness of the child deprived of its mother's milk by a new baby'. The wasting of muscles is masked by swellings caused by water accumulating in the tissues.

Lipids. Fats and oils (triglycerides) are sources of energy. They are broken down in digestion to fatty acids, monoglycerides and diglycerides. Some fatty acids are essential for normal health.

Mineral elements. Some mineral elements are needed in much greater amounts than others (see Figure 18.3 and Table 18.1). But a varied diet containing enough calcium and iron will usually provide enough of all the other elements.

More calcium is needed by children and adolescents than by adults, because calcium is being used in the formation of bones and teeth. Also, a woman needs extra calcium in pregnancy and when breast feeding, because calcium is being passed on to her baby. A shortage of calcium (or vitamin D, see later) in childhood results in a deficiency disease called rickets which is marked by poor growth and by deformation of the bones of the legs.

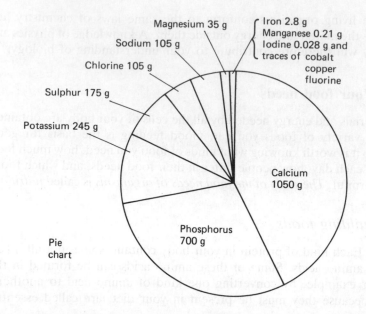

Magnesium 35 g

Sodium 105 g

Chlorine 105 g

Sulphur 175 g

Potassium 245 g

Iron 2.8 g
Manganese 0.21 g
Iodine 0.028 g and
traces of cobalt
copper
fluorine

Calcium
1050 g

Phosphorus
700 g

Pie
chart

Figure 18.3 mineral elements in your body

Table 18.1 *Some mineral elements and their use in your body*

Element	Use in body	Obtained from foods	Effect of deficiency in diet
Calcium	Calcium carbonate and calcium phosphate make bones and teeth hard	Cheese, milk, eggs and bread	Rickets in children, brittle bones in the elderly
Sodium	Sodium chloride in all body fluids	Enough added in food preparation	Cramps
Iron	Part of pigment that makes blood red	Liver, lean meat, wholemeal bread, green vegetables	Anaemia
Iodine	Part of thyroid hormones	Enough in most vegetables	Goitre in adults, dwarfism and mental retardation in children
Fluorine	Enamel production in teeth	Enough in most vegetables	Tooth decay

Iron is part of the haemoglobin molecule that makes blood red. Iron deficiency in the diet is one cause of anaemia. A woman needs extra iron in her diet to replace that lost in bleeding in her monthly periods. She also needs extra iron in pregnancy, for the blood of her baby and because the baby needs to store enough iron in its liver to last for some time after birth — until the baby starts to eat solid food.

Vitamins. With the development of chemistry in the nineteenth century, many foods were analysed and found to contain carbohdrates, lipids, proteins and mineral salts. But when in 1912 a British scientist, Gowland Hopkins, fed laboratory rats a mixture of these chemicals he was surprised to find that they did not grow properly. Other rats fed the same diet plus a little yeast or milk did grow normally. Hopkins concluded that yeast and milk contained some material that was essential to life. He did not know what it was but he called it a vitamin (Latin *vita* = life).

The vitamins were known first by letters, before they were chemically identified. They are now known to be organic molecules and have been given internationally agreed chemical names (see Table 18.2).

Some vitamins are gradually broken down in stored foods and are destroyed quickly when food is cooked. Foods containing vitamins are therefore best eaten fresh, and either raw or lightly cooked.

Apart from vitamin A, vitamins cannot be stored in the body. For good health you need small amounts in your diet every day. A temporary shortage of a vitamin in the diet may therefore cause a vitamin deficiency disease. (see Table 18.2). Note that rickets may be caused by a shortage of vitamin D as well as by a shortage of calcium.

Proteins, mineral elements and vitamins are essential structural materials for your body. Foods that contain these essentials are therefore called *body-building foods*. Because the essential mineral elements and vitamins are needed in small amounts, and because a shortage of any one of them results in a corresponding deficiency disease, foods containing these essentials are also called *protective foods*.

Energy is used whenever work is done in your body (see page 207). Foods that contain lipids (fats and oils) or carbohydrates (sweet foods, bread, potatoes, cassava and yams, for example) are called *energy-giving foods*. However, energy can also be obtained from proteins.

A balanced diet

You obtain from your food: (1) proteins, (2) carbohydrates, (3) lipids, (4) mineral salts, (5) vitamins, and (6) water. Your diet should also contain (7) roughage (indigestible fibres, especially cellulose from wholegrain cereals and fresh vegetables) which gives the diet bulk and helps to prevent constipation.

It is not easy to say how much you should eat each day. In general, a varied diet that provides enough energy will include the other essentials.

Table 18.2   *Vitamins, food sources and vitamin deficiency diseases*

Vitamin	Food sources	Deficiency diseases
Fat-soluble		
A Retinol	Carrots, butter, milk, eggs, liver, sardines and fish-liver oils	Slow growth in children, night blindness, infertility
D Calciferol	Formed in the skin in sunlight. Present in herrings, fish-liver oils, eggs and butter	Rickets in children, bone deformities in adults, (especially in old people)
E Tocopherol	Fresh green vegetables and eggs	Disorders of nerves and muscles
K Phylloquinone	Made by bacteria in the intestine but needed in the diet of young babies. Present in fresh green vegetables	Blood does not clot properly
Water-soluble		
B complex		
B_1 Thiamine	Wholegrain rice, wholemeal bread, milk, meat, yeast and potatoes	Check in growth of children, mental depression, beri-beri: swelling of legs, loss of appetite and paralysis
B_2 Riboflavine	Milk, cheese and liver	Check in growth of children, cracks in corners of mouth
Nicotinic acid (niacin)	Yeast, liver, beef, bread, fruit and vegetables	Check in growth of children, pellagra: skin rough red when exposed to light
B_{12} Cobalmin	Very lightly cooked liver. Also meat, milk, eggs and fish	Pernicious anaemia, degeneration of nerve cells
C Ascorbic acid	Citrus fruits, blackcurrants and fresh green vegetables	Check in growth of children, wounds do not heal, scurvy

This is why the daily requirement is usually expressed as the amount of energy required — measured in joules (see page 108).

A relatively inactive man who weighs about 86 kg needs about 12.7 MJ of energy each day. Of this about 5.5 MJ will be used for physical work and 7.2 MJ will be used whether or not he is active. This basal requirement is due to the use of energy by all cells even when a person is resting (for example, in cell maintenance, cell growth, cell division, digestive secretion, absorption of food, formation of food reserves, contraction of muscle, and the conduction of nerve impulses). However, the amount of energy-giving food you need each day depends on your size, and therefore on your age, and also on your occupation, the climate where you live, and your sex. Furthermore, there are individual differences that nobody understands.

Undernourishment and malnourishment

People who do not get enough to eat, day after day, are undernourished. Their food reserves are used up, then the structural materials of their body. Their health suffers and they are more prone to infectious diseases.

Even people who are not short of food may have a diet that is deficient in some essential. Such people, if they show deficiency symptoms, are said to be malnourished. Obesity is also a sign of malnourishment, when it is the result of overeating.

Some fungi and the parts of some flowering plants are poisonous. Even good food may be unfit to eat if contaminated by food-poisoning bacteria, or by some microscopic fungi. Never eat any food unless you are sure that it is edible and free from contamination. Also note that illness can result from an exceptionally large intake of some essentials (including common salt and vitamins A and D).

18.3 Your food canal

Your mouth is the opening into your mouth cavity, containing your teeth and tongue. *The intake of food* is called feeding or *ingestion*. As food is chewed it is mixed with saliva. This is the beginning of digestion — the physical and chemical breakdown of food.

A baby's teeth start to develop before birth and break through the skin during infancy and childhood. Your first set of teeth, called milk teeth, are replaced by a second set — called permanent teeth (see Figure 18.4). If you care for your teeth they could last a lifetime. Therefore, to keep your teeth healthy, clean them after each meal and do not eat sweets or snacks between meals.

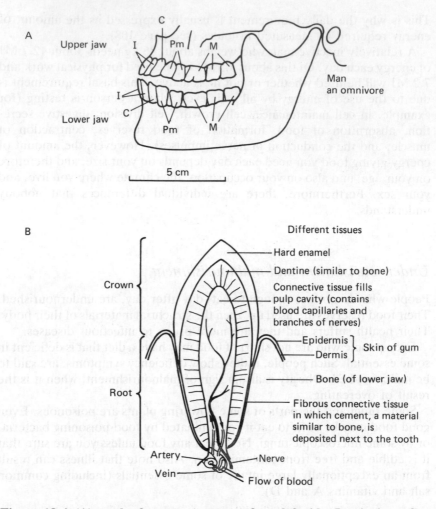

Figure 18.4 **(A) teeth of a man or woman from left side: I = incisor, C = canine, Pm = premolar, and M = molar. (B) Tissues in a tooth (vertical section): diagram not to scale. (From Barrass, R. Human Biology Made Simple)**

Digestion

Cells that produce materials which are used outside those cells are called gland cells. The useful materials passed out of a cell are called *secretions*, and their expulsion from the cell is called secretion. A group of secretory cells is called a gland. And the tube that carries a secretion from a gland, for example into the food canal, is called a duct.

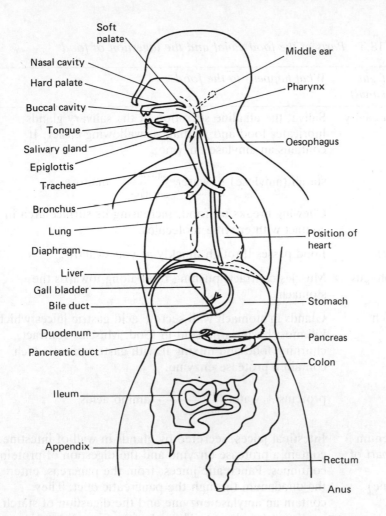

Figure 18.5 digestive system and gas exchange system. Diagram not to scale. Complete arrow = passage of food. Dotted arrow = passage of air. (From Barrass, R. Human Biology Made Simple)

Digestive secretions in different parts of your food canal or gut (see Figure 18.5 and Table 18.3) contain different enzymes. Each enzyme is a protein that acts as a catalyst. *A catalyst is a chemical that speeds up a chemical reaction: it takes part in the reaction but is liberated unchanged at the end of the reaction.* As a result, each molecule of the enzyme can be used repeatedly (see Figure 19.1).

Table 18.3   *Parts of the food canal and the digestion of food*

Part of gut (food canal)	What happens to the food
Mouth cavity	Saliva, the alkaline secretion of the salivary glands, lubricates food and so makes swallowing easier. It contains an amylase enzyme: $$\text{starch (amylose)} + \text{water} \xrightarrow{\text{salivary amylase}} \text{maltose}$$ Chewing breaks up food, increasing its surface area in contact with enzyme molecules.
Pharynx	Food passes through quickly as you swallow.
Oesophagus	Muscles contract, pushing food along towards the stomach.
Stomach	Glands in stomach wall secrete acid gastric juices which kill most bacteria present in food. Muscles contract, churning food and mixing it with gastric juices which contain a protease enzyme. $$\text{proteins} + \text{water} \xrightarrow{\text{gastric protease}} \text{amino acids}$$
Duodenum (first part of small intestine)	Intestinal juices, secreted by glands in wall of intestine, contain a protease enzyme and the digestion of proteins continues. Pancreatic juices, from the pancreas, enter the duodenum through the pancreatic duct. They contain an amylase enzyme and the digestion of starch continues (as above). They also contain a lipase enzyme: $$\text{triglycerides (lipids)} \xrightarrow{\text{pancreatic lipase}} \text{diglycerides} + \text{monoglycerides} + \text{fatty acids}$$ Bile produced in liver, and stored in gall bladder, enters duodenum through bile duct. Bile emulsifies fats and oils, increasing the surface area of these lipids in contact with the lipase enzyme. Both intestinal and pancreatic juices are alkaline. These neutralise acids from the stomach, and enzymes in the intestine work in an alkaline medium.

Note. Each enzyme catalyses a particular chemical reaction and enzymes are named according to the reaction they catalyse. For example, the chemical breakdown of starch (amylose) is a hydrolysis, a chemical combination with water (see Table 18.3). Contrast hydrolysis with a condensation reaction (see Figure 18.2). The hydrolytic enzymes involved in the digestion of starch (amylose) are called amylase enzymes. Proteases are involved in the digestion of proteins; and lipases in the digestion of lipids.

Absorption and assimilation

Some chemicals (including alcohol and some other drugs) are *absorbed* by cells lining the stomach. They pass into the blood and quickly affect other parts of the body.

The small food molecules resulting from the digestion of food macromolecules (see page 195 and Table 18.3) are *absorbed* by cells lining the small intestine (see Figure 19.2). *The incorporation of these materials into the materials of the body is called assimilation.* That is to say, once absorbed the materials become part of the body tissues or part of the body fluids. They are used in the great variety of chemical reactions involved in respiration, in cell maintenance and repair, and in cell growth and reproduction.

The digestive secretions contain water, and water is present in most foods as well as in drinks. Some of this water is absorbed in the small intestine, with the mineral salts and vitamins present in the food. Also, the molecules resulting from digestion are absorbed in solution in water. Then water is absorbed by cells lining the large intestine. As a result, the last part of the large intestine (the rectum, see Figure 18.5) may contain fairly dry faeces. These are the indigestible remains of the food eaten. *The passing out of faeces*, through the anus, is called defaecation or *egestion*.

Investigation 18.1 *The digestion of starch (amylose) by salivary amylase*

You will need a 0.5 per cent starch solution, iodine solution (see page 164), warm drinking water, a clean rubber band, two test tubes, two small plastic syringes and a spotting tile.

Warning. Organisms that cause disease may be present in the mouth cavity. So that there is no danger of passing on any infection from person to person in a laboratory use only clean materials provided by a teacher. Use only your own saliva, and do not put your lips to a test tube used by anyone else and wash up your own equipment afterwards.

Proceed as follows
1 Wash out your mouth with warm water.

3 Mix the saliva with a little warm water in your mouth and then collect about 4 cm³ of this mixture in a test tube.

4 Pour half of this into another test tube, boil it for 10 s then allow it to cool. Draw 1 cm³ into a clean syringe. Label this syringe: boiled saliva.

5 Draw 1 cm³ unboiled water/saliva mixture into the other clean syringe.

6 Draw 1 cm³ 0.5 per cent starch solution into each syringe.

7 Hold the syringes upside down to mix the contents. Then test the mixtures separately, at one minute intervals, by adding one drop to one drop of iodine solution on the spotting tile.

8 Keep a record of your observations. What do you conclude?

You may conclude that saliva contains something that facilitates the digestion of starch, but which is destroyed by boiling. All enzymes are proteins, and proteins are altered by boiling. This is why enzymes are ineffective as catalysts after they have been boiled.

19 The use of matter and energy in your body

The different kinds of molecule made by green plants (including carbohydrates, lipids, proteins and vitamins) are present in the food of animals. They are digested and smaller molecules are absorbed. Because of its length (about 7 m) and the presence of finger-like processes, the villi (see Figure 19.2), your small intestine has a very large surface area of epithelial cells absorbing the products of digestion. When molecules absorbed after digestion become part of the cells and tissue fluids of your body, they are said to have been assimilated.

19.1 Metabolism

In all living cells there are two kinds of chemical reactions: (1) *anabolic reactions* in which large molecules such as polysaccharides, lipids and proteins are made from smaller molecules (see page 195); and (2) *catabolic reactions* in which larger molecules are split into smaller molecules (as in respiration). Metabolism is the sum total of all chemical reactions in the cell (anabolic plus catabolic reactions).

Enzymes in cells

Many chemical reactions take place only if there is an input of energy. When the reaction takes place in a test tube energy may be provided by heating the reactants. In a living cell energy is available from respiration, and in the presence of appropriate enzymes reactions take place rapidly even at relatively low temperatures.

In plants, the sugars produced in photosynthesis are converted to starch (a reserve polysaccharide). The enzyme involved in this anabolic reaction is an amylase:

$$\text{maltose} \xrightarrow{\text{amylase}} \text{amylose} + \text{water}$$
$$\text{(a sugar)} \qquad\qquad \text{(starch)}$$

Starch, a reserve carbohydrate in many seeds, is converted to sugar during germination and used in respiration as an energy source.

$$\text{maltose} \xleftarrow{\text{amylase}} \text{amylose} + \text{water}$$

Note that this is a reversible reaction and the same enzyme is involved in the anabolic and catabolic reactions:

$$\text{maltose} \underset{}{\overset{\text{amylase}}{\rightleftharpoons}} \text{amylose} + \text{water}$$

Enzymes speed up chemical reactions (see Figure 19.1) and make the reactions possible at the low temperature of a living cell. But they do not affect the direction in which each reaction proceeds.

All enzymes that act as catalysts inside living cells are called intracellular enzymes. Other enzymes, which are secreted by cells and then active outside (for example, in your gut: see page 205) are called extracellular enzymes.

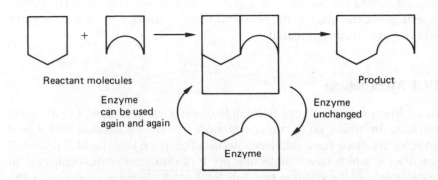

Reactant molecules

Enzyme can be used again and again

Enzyme

Product

Enzyme unchanged

Figure 19.1 the enzyme molecule provides a surface upon which the reactant molecules may become arranged close together and in correct alignment, aiding chemical combination (or the reverse reaction). (From Barrass, R. Human Biology Made Simple)

19.2   Your liver: its structure and functions

Your liver is a large lobed organ (see Figure 18.5) closely associated with your food canal. It is the only organ in your body that has a double blood supply: receiving blood from both the hepatic artery and the hepatic portal vein (see Figure 20.2).

Your stomach and intestines receive blood from a number of arteries but all this blood, after flowing through capillaries in the wall of the gut (see Figure 19.2A and B), flows into the hepatic portal vein and so to the liver (see Figure 20.2). Most of the food molecules absorbed by the epithelial cells of the small intestine (see Figure 19.2 A, B and C) pass into the blood and are transported directly to the liver.

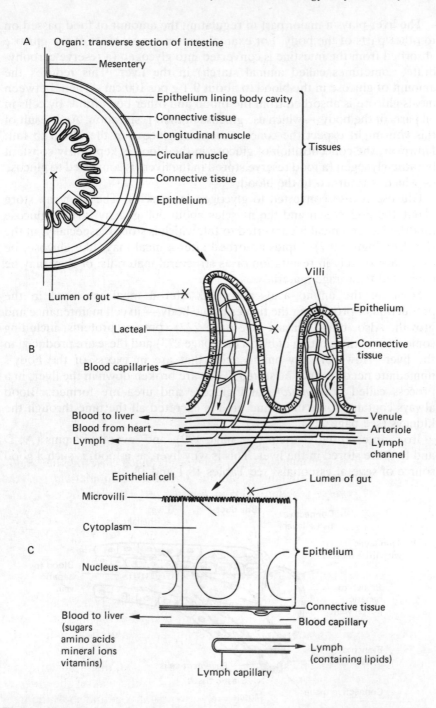

A   Organ: transverse section of intestine

— Mesentery

— Epithelium lining body cavity
— Connective tissue
— Longitudinal muscle
— Circular muscle   } Tissues
— Connective tissue
— Epithelium

Lumen of gut

B

Villi

Epithelium

Lacteal

Connective tissue

Blood capillaries

Blood to liver
Blood from heart
Lymph

Venule
Arteriole
Lymph channel

Epithelial cell

Lumen of gut

Microvilli

Cytoplasm

Epithelium

Nucleus

Connective tissue
Blood capillary

Blood to liver
(sugars
amino acids
mineral ions
vitamins)

Lymph
(containing lipids)

Lymph capillary

Figure 19.2 Small intestine: (A) half of a transverse section; (B) two villi; (C) epithelial cells with microvilli. Diagram not to scale.
(From Barrass, R. Human Biology Made Simple)

The liver plays a major part in regulating the amount of food passed on to other parts of the body. For example, after a meal most of the glucose absorbed from the intestine is converted into glycogen (a reserve carbohydrate, sometimes called animal starch) in the liver. This reduces the amount of glucose in the blood to about 0.1 g per 100 cm^3. Then, between meals glucose is absorbed from the blood and other body fluids by cells in all parts of the body — which use glucose in their respiration. As a result of this you might expect the concentration of glucose in the blood to fall. However, the concentration of glucose in the blood is kept fairly constant because glycogen (a food reserve stored in the liver) is converted to glucose — which is returned to the blood.

Glucose is also converted to glycogen in muscles. The liver can store about 100 g glycogen and the muscles about 300 g. Any further glucose absorbed after a meal is converted to fat, which is stored, especially, in the skin (see Figure 9.7). Lipids absorbed after a meal may, like glucose, be used immediately in respiration or as structural materials, or they may be converted to storage materials.

Some of the amino acids absorbed after a meal are used in the production of proteins in the tissues of the body — in cell maintenance and growth. Also, blood plasma (see page 222) contains proteins, including some used in the clotting of blood (see page 223) and these are produced in the liver. However, any amino acids that are in excess of the body's immediate needs cannot be stored. They are broken down in the liver, in a process called *deamination*, and glucose and urea are formed. Blood always contains some urea, and urea is excreted all the time through the kidneys (see page 217).

Iron, an essential element (see page 199), and several vitamins (A, D and B$_{12}$) are stored in the liver. This is why liver, as a food, is such a good source of several essentials (see Tables 18.1 and 18.2).

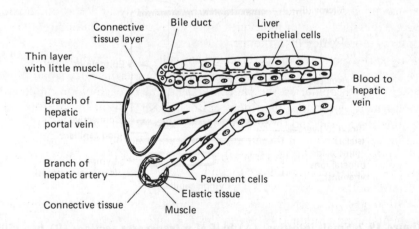

Figure 19.3 flow of blood and flow of bile in relation to liver epithelial cells. (From Barrass, R. Human Biology Made Simple)

The liver is a complex structure. It is concerned not only with regulating the composition of the blood but also with producing bile (see page 204). Figure 19.3 will help you understand that the liver's epithelial cells which carry out the regulatory functions are also the gland cells which produce bile.

Some toxic chemicals absorbed from the gut, including alcohol, are broken down into non-toxic molecules in the liver. However, alcohol does harm the liver. Regular drinkers, especially heavy drinkers, are likely to suffer liver damage, making this organ less effective in all its functions. Liver damage can cause death.

19.3 Your lungs and gas exchange

Oxygen is used and carbon dioxide is produced all the time in the respiration of your body cells. Any cell membrane through which oxygen can enter and carbon dioxide leave your body is called a *gas exchange membrane*. In unicellular organisms, the whole body surface is a gas exchange membrane; and in some multicellular organisms (including earthworms and frogs) gas exchange also takes place through all parts of the body surface. However, your skin is an effective barrier which restricts water loss and so hinders the diffusion of oxygen, and carbon dioxide molecules.

You have about 350 million air sacs in each lung (see Figure 19.4). Each one is very small but because there are so many their total surface area in contact with the air is very large (about 25 m^2 in each lung). This is the surface through which gas exchange between your body and the air takes place. Each air sac is formed of very flat pavement epithelial cells, and the surface membrane of each of these cells facing the air is a *gas exchange membrane* though which oxygen enters your body and carbon dioxide leaves. Inevitably, water molecules are also lost. So the air you breathe out contains more water than the air you breathe in.

The surface of each of these pavement cells that is furthest away from the air (see Figure 19.4) is closely applied to numerous very small blood vessels (blood capillaries, see page 221). The wall of each blood capillary is also formed of very flat pavement epithelial cells. Because both the cells forming the air sac and those forming the blood capillaries are so flat, the distance from the air in the air sac to the blood in the blood capillaries is very short. Oxygen diffuses from the air into the blood and carbon dioxide diffuses from the blood into the air — over this *very short diffusion distance*.

Differences in oxygen and carbon dioxide concentrations (diffusion gradients) are maintained. *There is always a higher concentration of oxygen and a lower concentration of carbon dioxide in the air sacs than in the blood*, because:

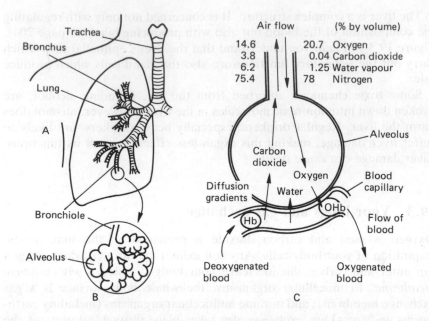

Air flow	(% by volume)	
14.6	20.7	Oxygen
3.8	0.04	Carbon dioxide
6.2	1.25	Water vapour
75.4	78	Nitrogen

Figure 19.4 gas exchange: (A) part of gas exchange system; (B) air sacs at end of a bronchiole; (C) gas exchange between blood in a blood capillary and the air in an air sac. Hb = haemoglobin; OHb = oxyhaemoglobin. (From Barrass, R. Human Biology Made Simple)

1 The oxygen that diffuses into the blood combines with the red blood pigment haemoglobin in the erythrocytes and is carried away from the lungs. This blood from the lungs goes to other body tissues (see Figure 20.2.) before returning to the lungs.
2 In all cells of the body, oxygen is being used all the time in aerobic respiration and carbon dioxide is being produced. As a result, oxygen diffuses out of the blood and carbon dioxide diffuses into the blood in all organs except the lungs: that is to say the diffusion gradients are reversed.
3 As a result of losing oxygen and gaining carbon dioxide in other parts of the body, the blood that flows back to the lungs contains much less oxygen and much more carbon dioxide than when it left the lungs.
4 Breathing (see page 122), ventilates the lungs and the air taken into the lungs always contains more oxygen than the blood and much less carbon dioxide.

Note. In everyday language, some people say respiration when they mean breathing, and when someone stops breathing he or she may be given artificial respiration. To avoid confusion, in science, the word *breathing* is used for the repeated filling and emptying of the lungs, *ventilation* for the flow of air in and out of the lungs, and *gas exchange* for the diffusion that occurs across membranes. The word *respiration* should be restricted to the controlled release of energy from organic molecules, including sugars (see page 168), that occurs all the time in all living cells.

Breathing polluted air

Some chemicals in air, including sulphur dioxide produced when coal is burned, irritate and damage cells of the air passages and are probably an indirect cause of bronchitis. Carbon monoxide in car exhaust fumes makes the haemoglobin in blood ineffective and is most dangerous in confined spaces, such as garages, where people will die quickly from carbon monoxide poisoning if a car engine is left running.

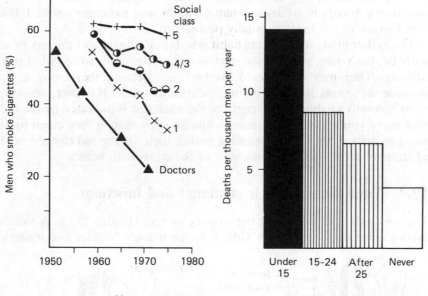

Figure 19.5 graph: giving up smoking: changes in British society. Percentage men (aged 16+) in each social class (1 to 5) who smoked cigarettes, and for comparison percentage of male doctors (aged 40+). (Based on Smoking or Health **3rd Report of Royal College of Physicians)**
histogram: dying for a smoke: number of deaths each year per 1000 men (aged 45 to 54) in the USA. The earlier a man starts to smoke the more likely is he to die before the age of 54. (Based on Hammond, E. C. National Cancer Institute Monograph, **19)**

Glues and cleaning fluids contain solvents that depress brain activity when the vapours are inhaled and can cause death. Great care is taken in factories where these products are made.

Try to avoid breathing fumes and vapours into your lungs. If you smoke tobacco, tar is deposited in the air passages of your lungs, making them less efficient. Nicotine, to which you quickly become addicted (see page 231), and other harmful chemicals, are absorbed directly into your blood. By smoking, you increase your chances of an early death from bronchitis, lung cancer, or heart disease (see Figure 19.5). Even non-smokers are affected if they breathe smoke-laden air — this is called passive smoking.

The risk of death from lung cancer is related to the number of cigarettes smoked each day and to the age when smoking started. Under the age of 65, smokers are about twice as likely as non-smokers to die of coronary heart disease (heart attack). Smokers are also more likely than non-smokers to suffer from bronchitis.

Ten years after giving up smoking, a person is no more at risk from lung cancer or coronary heart disease than someone who has never smoked. But any damage to the lungs is usually permanent.

The earlier in life the smoking habit is acquired, the greater the risk of an early death. Young people may start to smoke cigarettes to see what they are like. They may continue to smoke – not because they enjoy it but because they think it makes them appear more adult. However, smoking soon becomes a habit. The strength of the addiction is indicated by the fact that many young people continue to smoke even though they claim to be short of money, they know smoking makes their breath and clothes smell of stale tobacco, and they are aware of the dangers to health.

19.4   Your kidneys: their structure and functions

The two kidneys are part of the urinary system (Figure 19.6). A ureter carries urine away from each kidney to the urinary bladder and a single

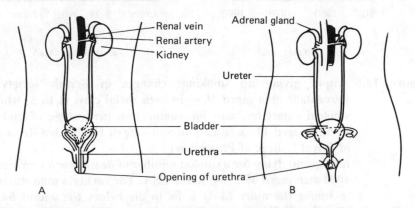

Figure 19.6 urinary system: (A) of a man; and (B) of a woman. (From Barrass, R. Human Biology Made Simple)

tube, the urethra, carries the urine from the bladder to the outside when the muscles in the wall of the bladder contract. A renal artery carries blood into each kidney and a renal vein carries blood away.

Investigation 19.1   *Dissection of a sheep's kidney*

You will need a fresh kidney from a butcher's shop. If possible obtain one with the ureter, renal artery and renal vein cut a few centimetres away from the kidney.

Proceed as follows. Cut the kidney in half so that you can see the parts labelled in Figure 19.7A. If possible, look at a demonstration dissection of a small mammal (Investigation 5.5) and note the position of its kidneys, ureters, bladder and urethra (also note its liver and lungs). Wash your hands and the work area thoroughly afterwards.

In each kidney, the blood capillaries are closely associated with renal tubules. There are about a million renal tubules in each of your kidneys

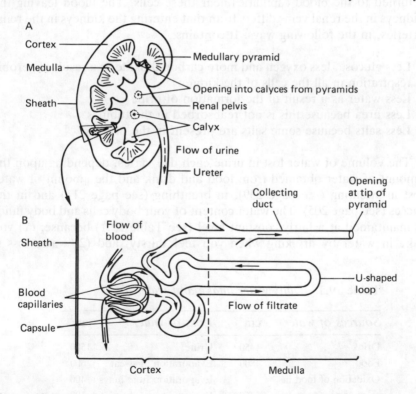

Figure 19.7 structure of a kidney: (A) kidney cut in half; (B) the parts of one renal tubule. (From Barrass, R. Human Biology Made Simple)

though only one is shown in Figure 19.7B. The parts of a renal tubule are the capsule, the first convoluted tubule, the U-shaped loop and the second convoluted tubule which opens into a collecting duct. Note (in Figure 19.7A) that the capsule and the two convoluted tubules are in the outer part of the kidney (the cortex) and the U-shaped loop and collecting duct are in the inner part (the medulla).

Loops of blood capillaries are closely associated with the cup-shaped end of each renal tubule (see Figure 19.7B). The blood in the capillary loops is under high pressure and some of the blood (the smaller molecules including water, urea and glucose) filters through into the renal capsule.

The composition of the filtrate changes as it passes along the tubule because the cells that form the wall of the tubule absorb some materials (for example most of the water and all of the glucose) but not others (for example, urea — see page 210). Mineral salts are selectively reabsorbed. As a result, the urine that eventually flows from each kidney, in the ureter, is very different in composition from the filtrate forced into the renal capsules.

Materials absorbed by the cells of the renal tubules, from the filtrate, are returned to the blood capillaries near these cells. The blood leaving the kidneys in the renal veins differs from that entering the kidneys in the renal arteries, in the following ways. It contains:

1 Less glucose, less oxygen and more carbon dioxide as a result of aerobic respiration in all the cells of the kidney
2 Less water as a result of the formation of urine
3 Less urea because this is not reabsorbed in the tubules
4 Less salts because some salts are present in the urine.

The volume of water lost in urine each day varies, depending upon the amount of water obtained from food and drink and the amount of water lost in sweating (see page 239), in breathing (see page 211) and in the faeces (see page 205). The water content of your body cells and body fluids is maintained at a fairly constant level (see Table 19.1) because: (1) you take in water by drinking when you feel thirsty, and (2) water loss is

Table 19.1   *Daily water intake and water loss*

Sources of water	cm^3	Loss of water	cm^3
Drink	1450	Urine	1500
Food	800	Evaporation of sweat	600
Oxidation of food in		Evaporation from lungs	400
respiration	350	Faeces	100
Total	2600		2600

regulated by behaviour (including resting in the shade on hot sunny days) and by the involuntary control of the amount of urine produced by your kidneys. *The control of the water content of the body,* as a result of the balance between water intake and water loss, is called *osmoregulation.*

Because they play a part in the elimination of the waste products of metabolism (for example, urea), the kidneys are also excretory organs. Because salts are selectively reabsorbed from the filtrate in the renal tubules — some are lost and others retained — the kidneys also help maintain the concentration of salts in the cells and fluids of the body at a fairly constant level.

Excretion is the *elimination of waste substances produced in the cell* (waste products of metabolism, see page 207) *and of other substances that are present in excess in the cells and tissue fluids.* These would be toxic (poisonous) if they accumulated in the body. The loss of carbon dioxide, a waste product of respiration in all body cells, occurs through the lungs and is called carbonaceous excretion. The loss of urea, a nitrogen-containing compound that is a waste product of the breakdown of amino acids in the liver, occurs through the kidneys and is called nitrogenous excretion. See also pages 209 and 216.

Excretion should not be confused with egestion (the elimination of faeces, see page 205). These technical terms are used by scientists for two different processes which are commonly confused in everyday language. And excretion should not be confused with secretion (in which useful materials are passed out of a cell, see page 205). But note that in the secretion of sweat (which plays a part in temperature control) there is a loss of sodium chloride (0.3 per cent) and some urea (0.03 per cent). This loss from the body is called uncontrolled excretion because the volume of sweat produced varies from time to time and is not related to the amount of water, salts or urea in the body.

People who lose a lot of water by sweating must replace the salt as well as the water. To prevent dehydration — the excessive loss of water that can cause death — water lost in vomit, in diarrhoea, and in sweating during a fever, must be replaced by drinking large quantities of clean water.

⬡20 **Your body: parts working together**

20.1 Blood and other body fluids

The circulation of blood

Your circulatory system comprises your heart and a system of tubes in which blood flows from the heart to all tissues of the body, and from these tissues back to the heart. It is called the circulatory system because, as a result of the repeated contractions of the heart, blood circulates through these tubes throughout life. This fact was first demonstrated by a British physician, William Harvey, whose book *On the Anatomy and Motion of the Heart and Blood in Animals* was published in 1628. Before then, people believed that blood from the heart ebbed and flowed in the veins.

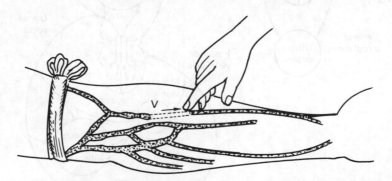

Figure 20.1 the directional flow of blood in veins: one of Harvey's demonstrations. When blood is pressed out of one of the superficial veins of the forearm, as indicated by the arrow, the valve (at V) prevents back flow. (From Barrass, R. Human Biology Made Simple)

Investigation 20.1   *The directional flow of blood*

You will need only a bandage, to enable you to repeat Harvey's demonstration.

Proceed as follows

1 Tie the bandage around your arm, just below the elbow (as in Figure 20.1).

   Warning. Do not keep the bandage tight for more than one minute.

2 Stroke one of the veins, pressing towards your wrist (as in Figure 20.1). The backflow of blood, as far as a valve in the vein, will cause the vein to bulge. You may conclude that the valves in the veins allow the blood to flow in one direction only — away from your wrist, towards your heart.

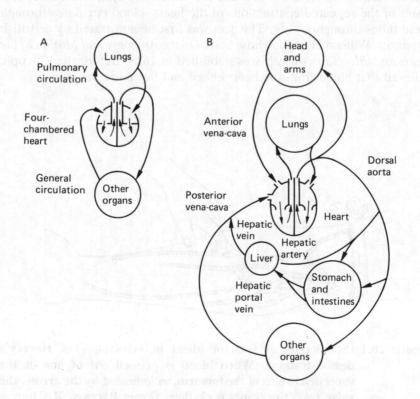

Figure 20.2 circulatory system of a mammal: (A) pulmonary and general circulation; (B) position of liver in general circulation. (From Barrass, R. Human Biology Made Simple)

In the body of a mammal there is a *double circulation*, shown in Figure 20.2A as a figure of eight. In the *pulmonary circulation* blood flows from the right side of the heart, through the lungs, to the left side of the heart. And in the *general or systemic circulation* blood flows from the left side of the heart, through all other organ systems, to the right side of the heart. The heart is a four-chambered muscular pump.

Arteries carry blood from the heart to all organs of the body. They have thick, elastic, muscular walls. There is one artery to each organ, and branches within each organ that carry the blood to the capillaries.

Blood capillaries are numerous thin-walled tubes which carry blood close to all living cells. They rejoin to form small veins, the tributaries of larger veins.

Veins carry blood from the tissues back to the heart. Their walls are thinner than those of arteries, and they have valves which allow blood to flow in only one direction — towards the heart.

Note. The liver is the only organ that receives a blood supply from a vein as well as an artery. (see page 208). All arteries except the pulmonary artery carry oxygenated blood, and all veins except the pulmonary vein carry deoxygenated blood. (see page 212).

The heart

Blood pours into the heart from the posterior vena cava and from two anterior vena cavae. These are the largest veins. (Only one anterior vena cavae is shown in Figures 20.2 and 20.3).

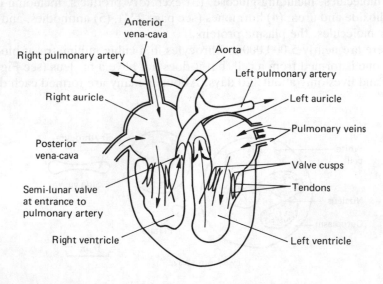

Figure 20.3 the heart of a mammal. Diagram not to scale. (From Barrass, R. Human Biology Made Simple)

The right auricle has a thin muscular wall. When this contracts blood is pumped into the right ventricle. The valve between the auricle and the ventricle prevents any back flow of blood.

The right ventricle has a thicker, more muscular wall. When this contracts blood is pumped along the pulmonary arteries, through the capillaries of the lungs, along the pulmonary veins, and into the left auricle. Valves at the entrance to the pulmonary artery, and in the veins, prevent any back flow of blood.

The left auricle has a thin muscular wall. When this contracts blood is pumped into the left ventricle. The valve between the auricle and the ventricle prevents any back flow of blood.

The left ventricle has a very thick muscular wall. When this contracts blood is forced into the aorta and so to all organs of the body (except the lungs). Valves at the entrance to the aorta prevent any back flow of blood. Your heart, like all other organs, has a blood supply. The coronary arteries, which branch from the aorta next to the heart, carry blood to the muscles of the heart.

The heart beats repeatedly, throughout life. Because of the elasticity of the arteries and the resistance offered by the capillaries to blood flow, the arteries remain full of blood at all times — even between heart beats.

Blood, tissue fluid and lymph

The fluid part of blood is called blood plasma. In this fluid are suspended erythrocytes, white blood cells and blood platelets. Blood plasma is mostly water but it contains, in solution: (1) salts, including sodium chloride, (2) food molecules, including glucose, (3) excretory products, including carbon dioxide and urea, (4) hormones (see page 224), (5) antibodies, and (6) larger molecules, the plasma proteins.

There are nearly 5 000 000 erythrocytes in a cubic millimetre of blood. Each one is formed from a cell, but it does not have a nucleus (see Figure 20.4) and lives only about 120 days. About as many are formed each day,

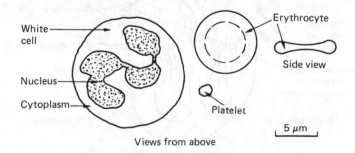

Views from above

Figure 20.4 a white cell, an erythrocyte, and a blood platelet. (From Barrass, R. Human Biology Made Simple)

in the red bone marrow of some bones, as are destroyed in the liver. Because of its shape (a biconcave disc, see Figure 20.4), an erythrocyte has a very large surface area in proportion to its volume. It contains the red pigment haemoglobin, which accepts oxygen in regions of high oxygen concentration (from the air sacs of the lungs) and gives up oxygen in places where the oxygen concentration is lower (in all other body tissues).

There are about 5000 white cells in a cubic millimetre of blood (see Figure 20.4). They feed on bacteria and so help to prevent and cure diseases. They also produce: (1) chemicals called anti-toxins which neutralise the toxic chemicals (toxins) produced by bacteria, and (2) antibodies which help to destroy bacteria.

A cubic millimetre of blood contains about 300 000 blood platelets (see Figure 20.4), which disintegrate in damaged tissues, releasing an enzyme that catalyses the first of a sequence of reactions involved in the clotting of blood. In the last of these reactions a soluble blood protein, fibrinogen, is converted to threads of fibrin which form a mesh that plugs the wound. Cells next to the wound divide, producing new cells of different types that replace the damaged cells.

In all tissues of the body some of the blood plasma (water with ions and small molecules in solution but not the large plasma protein molecules) passes through the walls of the blood capillaries and bathes nearby cells. These cells extract food molecules and absorb oxygen from this tissue fluid, and excrete carbon dioxide and other waste products of their metabolism into it.

As a result of diffusion gradients between blood and the tissue cells, oxygen diffuses from the blood into the tissue fluid and carbon dioxide diffuses from the tissue fluid into the blood (see page 212).

Some water is drawn back into the blood capillaries (see Figure 20.5) and some tissue fluid (which now contains fewer food molecules but more

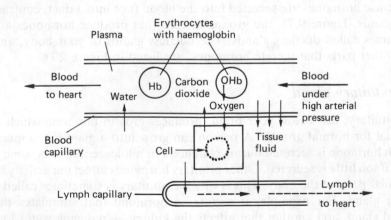

Figure 20.5 the relationship between blood, tissue fluid and lymph. (From Barrass, R. Human Biology Made Simple)

excretory products) passes into the lymph capillaries (see Figure 20.5). Lymph capillaries in all tissues (see lacteal, Figure 19.2) carry their fluid contents, called lymph, towards larger lymph vessels which have valves that prevent back flow. The largest lymph vessels drain into the large veins near to the heart. This system of one-way capillaries and vessels is called the lymphatic system. As a result of the drainage away of fluid in the lymphatic system, tissue fluid does not normally accumulate.

The small amount of tissue fluid that permeates the tissues provides a watery environment, a mixture of almost constant composition, in which all cells live.

Functions of blood

1 Blood carries food materials to all parts of the body.
2 White cells help to combat diseases caused by bacteria.
3 Blood transports oxygen and carbon dioxide.
4 Blood transports urea, a waste product of protein metabolism.
5 The clotting of blood is the first stage in wound healing.
6 Blood transports hormones (chemical messengers).
7 Blood helps to maintain body temperature at a fairly constant level, by distributing heat throughout the body.

20.2   Chemical control: hormones

Activities in your body are coordinated, partly as a result of signals carried in nerves and partly by hormones (chemical messengers) carried in the blood. *A hormone is a chemical produced by cells in one part of the body, which is transported in the blood, and which affects cells in another part of the body that are particularly sensitive to it.*

Because hormones are secreted into the blood (not into a duct: contrast with glands, Figure 9.7), the groups of cells that produce hormones are sometimes called ductless glands. The ductless glands of your body, and some other parts that secrete hormones, are listed in Figure 20.6.

The pituitary gland

The pituitary, just below your brain, produces growth hormone which is essential for normal growth. A person can grow into a giant if too much growth hormone is secreted during childhood or adolescence, or become a dwarf if too little is secreted. Other pituitary hormones affect the activity of particular parts of the body. This is why the pituitary is sometimes called a master gland. For example, it secretes a hormone that stimulates the thyroid gland, and another that affects the kidney — reducing water loss when the water content of the body falls to a certain level. Other pituitary

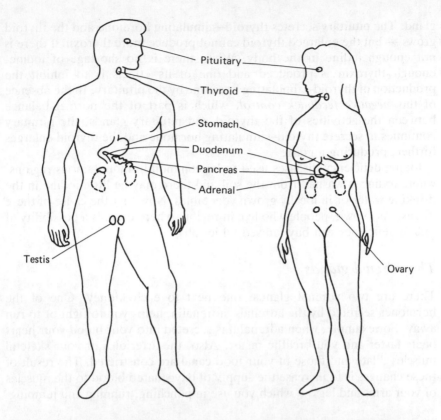

Figure 20.6 some parts of the body that secrete hormones. Diagram not to scale. (From Barrass, R. Human Biology Made Simple)

hormones, the gonadotropic hormones, stimulate the gonads (the testes of the male and the ovaries of the female).

The thyroid gland

The thyroid, between the voice box (larynx) and the skin of the neck, produces a hormone called thyroxin which stimulates respiration in all body cells. As a result, anyone who has an overactive thyroid may be thin, restless and highly strung; and anyone with an underactive thyroid is likely to be overweight and sluggish.

Thyroxin contains iodine. If there is a shortage of iodine in the diet the thyroid cannot produce enough thyroxin. Because thyroxin stimulates the secretion of growth hormone by the pituitary, children who do not have enough iodine in their diet do not grow properly. Iodine deficiency also results in goitre, a disease that is marked by the enlargement of the thyroid gland. The pituitary secretes thyroid–stimulating hormone and the thyroid grows — but the enlarged thyroid cannot produce more thyroxin if there is not enough iodine in the body. When there is no shortage of iodine, enough thyroxin is produced and one of its effects is to inhibit the production of thyroid–stimulating hormone by the pituitary. In the absence of this *negative feedback control*, which is part of the normal balance between the activities of the thyroid and pituitary glands, the pituitary continues to secrete thyroid–stimulating hormone and the thyroid enlarges further, producing a goitre.

Iodine deficiency diseases used to be common in mountainous regions, where iodine is washed from the soil and there may be little iodine in the drinking water or in locally grown vegetables. Now that the cause of these diseases is known, people who live in regions where there is a possibility of iodine deficiency can buy iodised table salt.

The adrenal glands

There are two adrenal glands, one next to each kidney. One of the hormones secreted by the adrenals, adrenalin, helps you to fight or to run away. For example, when adrenalin is secreted into your blood your heart beats faster and you breathe faster. Also, the arterioles of your skeletal muscles dilate and those of your food canal are constricted. The result of these changes is to increase the supply of oxygenated blood to the muscles of your arms and legs — which you use in punching, running and jumping.

The pancreas

The pancreas produces two kinds of secretion: (1) pancreatic juices pass along the pancreatic duct into the duodenum (the first part of the small intestine: see Figure 18.5), (2) insulin, a hormone, is secreted into the blood by small groups of cells called islets which are not connected to the pancreatic duct.

Insulin is necessary for the conversion of glucose to glycogen in the liver (see page 210). People who do not have enough insulin suffer from sugar diabetes. Any excess glucose is removed from their blood in their kidneys and is present in their urine. People who suffer from this disease must control their diet carefully. They may also be given insulin (by injection) so their blood sugar can be maintained at the normal level (0.1 g glucose per 100 cm^3 blood).

20.3   Neural control: nerves and nerve impulses

Your nervous system comprises your brain and spinal cord (your central nervous system) and the nerves which are links between your brain and spinal cord and other parts of your body. Long nerve cells called *neurones* (See Figure 9.6) carry signals (nerve impulses) and provide a link between *sense cells* (which are sensitive to stimuli) and *effector cells* (which are either muscles that contract or glands that secrete).

A nerve impulse is a signal that passes along a neurone but we do not know how this signal is transmitted. However, the transmission of a nerve impulse is an active process in which energy is used and which is associated with chemical and electrical changes. It is much slower than a current of electricity passing along a wire.

Some activities are automatically controlled. We make responses without thinking. Indeed, because we cannot prevent ourselves from making these responses, they are called *reflex actions*.

Investigation 20.2 *A cranial reflex*

You will need a mirror and a light source, such as a torch.

Proceed as follows
1 Look at one of your eyes in the mirror so that you can see the size of the pupil — through which light enters the eye.
2 Shine a light into the mirror, and note that your pupil almost immediately gets smaller.

Rapid changes in the size of your pupil in response to changes in light intensity are the result of a cranial reflex in which nerve impulses are carried from your eye to your brain and from your brain to your eye.

Activities controlled by your spinal cord, in which your brain is not directly involved, are called spinal reflexes. For example, cross your legs and then strike the tendon immediately below your kneecap. Your foot will jerk forward. This is called the knee-jerk response. The arrangement of neurones involved is shown in Figure 20.7. There are stretch receptors (sensors) in the muscle, which are sensitive to the slight pull on the tendon (a stimulus). The nerve cell that carries nerve impulses from a sensor into the spinal cord is called a *sensory neurone*. This is a very long nerve cell and its cell body is in the dorsal root ganglion, a swelling in the spinal nerve just outside the cord (see Figure 20.7). The nerve cell that carries nerve impulses away from the spinal cord to an effector (a muscle in the upper part of your leg) is called an *effector neurone*.

The gaps between the sensors and the sensory neurone, between the sensory neurone and effector neurone, and between the effector neurone and the effector, are called *synapses*. A chemical transmitter substance secreted by one cell at a synapse stimulates the transmission of an impulse in the next cell, or the contraction of the muscle fibre.

The knee-jerk reflex is an example of a response to stimuli in which information in the body is transported not by the distribution of a hormone in the blood but by the passage of nerve impulses in neurones. The sensory and effector neurones are a link between the sensor and the effector.

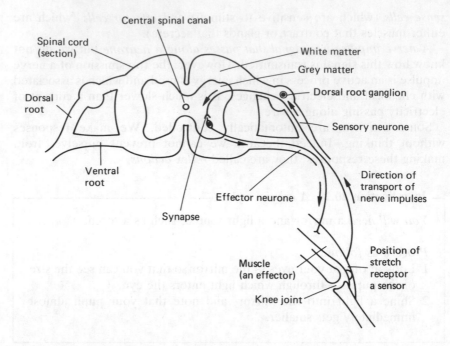

Figure 20.7 spinal reflex arc. Diagram not to scale. (From Barrass, R. Human Biology Made Simple)

Together they form what is called a reflex arc, with the sensory neurone carrying the nerve impulses from a sensor into the central nervous system (the coordinator) and the effector neurone carrying nerve impulses to an effector. By way of such reflex arcs, stretch receptors provide information about the state of muscle contraction – which makes possible both the maintenance of posture and the control of movement.

Some differences between hormonal and neural control

1 Hormones are carried in the blood and are therefore present in the tissue fluids that bathe every body cell. Their action depends upon certain cells being affected by them (target cells) while most cells are not affected. Nerve impulses, in contrast, are carried directly from the receptor to an appropriate effector.
2 The reaction to most hormones is automatic, like the reflex responses controlled by nerves. Hormones are not involved in the conscious control of our actions. This conscious control, mainly control of skeletal muscles, is made possible by nerve cells.
3 Hormonal control is quite rapid because hormones are carried in the blood, from the secretory cell to the target tissue, in the time it takes for

blood to circulate around the body (less than 30 s). But neural control is much faster. A nerve impulse travels at about 100 metres per second.

4 Hormones also control long-term changes (such as the 28-day menstrual cycle in women, see page 247) and changes that take several years (such as the development of the body to sexual maturity during adolescence).

Your brain

The brain is directly involved in all cranial reflex responses (for example, control of the size of the pupil in response to changes in light intensity, see page 227). Although the brain is not directly involved in spinal reflexes, other neurones (called connector neurones) convey information to the brain. You remove your finger when it is pricked with a pin. But as well as this spinal reflex response you learn not to prick your finger again.

Learning is the ability to benefit from experience; and intelligence is the ability to combine isolated experiences and draw correct conclusions. Your brain is involved in all voluntary activities — in which you decide what to do. As a result of the analysis of information by the brain, you usually behave in an appropriate way in any situation.

Scientists have discovered, for example by studying the behaviour of people suffering from brain damage, which parts of the brain are concerned with different activities (see Figure 20.8).

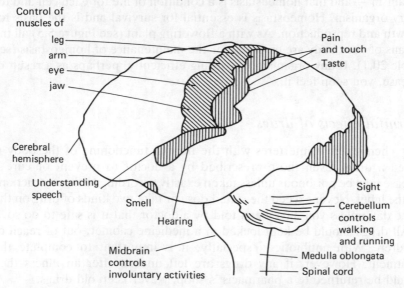

Figure 20.8 left side of the brain. Areas of cerebral cortex that receive nerve impulses from sense organs are labelled, as are those concerned with the control of muscles in different parts of the body. Other areas are concerned with thinking, understanding and remembering. (From Barrass, R. Human Biology Made Simple)

20.4 Homeostasis: constancy based on change

If a non–living thing such as this book is essentially the same this week as it was last week, then this is because it has not changed during the week. In contrast, if you are much the same as you were last week it is because you have been changing all the time. You have been taking energy and materials from your surroundings and losing energy and materials to your surroundings. Many cells in your body have been destroyed and many new cells have been formed. The circulation of your blood, linking all parts of your body, and the activity of the separate parts have maintained the constant composition of your tissue fluids (see page 223). The flow of lymph in your lymphatic system has prevented the swelling of your body tissues with tissue fluid. Claude Bernard, a French physiologist, wrote in 1857 that *the constancy of the internal medium is the condition of life.*

This steady state (the condition of constancy in any cell, in the body fluids and in the body as a whole) that results from continuous change, is called homeostasis (Greek *homoios* = similar; *stasis* = standing); and the processes that contribute to the maintenance of this steady state are called *homeostatic mechanisms*.

The working together of all parts of your body, made possible by hormonal and neural control, results in the maintenance of homeostasis. Remember that coordination and control are features of the life of every organism — and that homeostasis is a condition of life for each cell and for every organism. Homeostasis is essential for survival and is the basis for growth and reproduction. As with a flowering plant (see Figure 5.5) all the organs of your body are essential for the maintenance of homeostasis (see Table 20.1). If any organ stops working efficiently, perhaps as a result of disease, you soon feel ill.

Harmful effects of drugs

Any chemical that interferes with the normal functioning of the body is called a drug. Even those prescribed by a doctor to prevent or cure a disease may be poisonous unless taken exactly according to the instructions on the label. Never exceed the stated dose or take two kinds of drug on the same day unless you have been told by a doctor that it is safe to do so.

All drugs should be kept locked in a medicine cabinet, out of reach of children. With antibiotics, especially, it is important to complete the treatment. However, if any drugs are left unused after an illness they should be returned to a pharmacist's shop. Never keep old drugs.

Never take anyone else's drugs or allow anyone to take drugs prescribed for you. Especially, never give drugs prescribed for an adult to a child. If you are pregnant, or if there is a possibility that you are, never take any drug without first checking with your doctor that it is safe for you and for your baby. Remember, any drug will enter not only your own blood but also that of your baby.

Table 20.1 *Interdependence of parts of your body*

Part of organism	Some functions
Skeleton (skeletal system)	Maintenance of posture, protection and movement
Skeletal muscles (muscle system)	Support, maintenance of posture and movement
Food canal (digestive system)	Ingestion, digestion, absorption and egestion
Liver	*See text*
Lungs, bronchi, trachea (gas exchange system)	Intake of oxygen and excretion of carbon dioxide
Kidneys, ureters, bladder (urinary system)	Control of water and salt content of body and nitrogenous excretion
Heart, blood vessels (circulatory system)	Circulation of blood through all parts of body
Skin	Support, protection, control of heat loss, and sensitivity
Sense organs	Sensitivity (reception of stimuli)
Nervous system	Control of many activities, memory
Reproductive system	Production of young

Some drugs doctors do not prescribe

Alcohol is a drug that affects the brain immediately, slowing responses to danger and causing a loss of self-control. Under its influence people are more likely to have accidents; and may do things they regret for the rest of their lives. Too much alcohol can cause liver damage, brain damage and mental illness. Too much taken in too short a time can cause a coma or even death. Also, it is dangerous to take some drugs, including some taken as medicines, before or after alcohol.

Some people who drink alcohol soon find they cannot manage without this drug. This happens with many drugs and is called addiction. Nicotine (see page 214) is a drug to which tobacco smokers soon become addicted. This is why people who do not smoke never feel the need for a cigarette.

Most other drugs to which people may become addicted are available only from drug-pushers. There is no control over their strength or purity and their effect on the body is unpredictable.

Many addicts spend more money than they can afford on alcohol, nicotine or other drugs. As a result they, and perhaps their dependents, may go short of food, clothing or other necessities.

The only way to avoid becoming dependent on any drug, or being damaged by drug taking, is to refuse any drug that is offered. Take an interest in other things, and make friends with people who share these other interests.

The movement of heat and temperature control

21.1 Heat energy transfer

Heat energy is transferred by conduction, convection and radiation.

Conduction

If you hold one end of a solid rod and heat the other end, the rod may soon become too hot to hold. The heat energy causes molecules at the heated end to vibrate violently, and each vibrating molecule causes nearby molecules to vibrate. In this way heat travels slowly through most solids. This method of heat energy transfer is called conduction. Metals are particularly good *conductors of heat* (and of electricity, see page 158) because they have many free electrons. These can move, transferring kinetic energy rapidly from warmer to colder parts of the metal.

Non–metals are bad conductors of heat, and are called *heat insulators*. A pan is made of metal, which conducts heat rapidly, but the handle is made of wood or plastic which do not get too hot to hold, even when the pan contains boiling food.

Apart from liquid metals (for example, mercury and molten iron), all fluids (liquids and gases) are poor conductors of heat.

Convection

When fluids are heated they expand and so become less dense (see page 36). As a result, heated fluids (gases and liquids) rise, and cooler fluids fall and take the place of the heated fluids. This is why there are currents of fluid, as hotter and cooler fluids circulate (for example, see winds, page 269).

Investigation 21.1 *Observing currents in a liquid*

You will need the materials illustrated in Figure 21.1

Proceed as follows
Arrange the apparatus as in Figure 21.1 and heat the water gently on one side of the beaker. Observe the circulation of coloured water as heated water rises on one side of the beaker, cools, and falls on the other side. Keep a record of your observations in your note book.

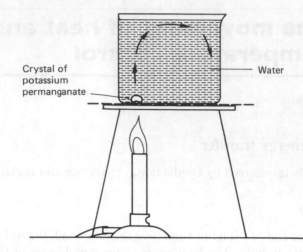

Crystal of potassium permanganate

Water

Figure 21.1 convection currents in a liquid

Convection is the transfer of heat energy, within a fluid, by the circulation of the fluid due to temperature differences. The currents of moving molecules are called convection currents. Convection occurs only in fluids, because the molecules in solids are not free to move.

Because air is a poor conductor of heat, it can be used as an insulator — to reduce heat loss. Compared with other insulators, it has the advantage of being free. However, when air is used as an insulator, care must be taken to minimise its movement, because heat travels through air by convection. For example, clothing for cold weather should be loosely woven or made of loosely packed material so that air is trapped near the body. To keep warm in cold weather you should also wear a windproof outer garment so that warm trapped air is not blown away.

Radiation

All hot objects give off energy in the form of radiant heat. This is similar to light energy but is from the invisible infrared part of the electromagnetic spectrum. Using infrared–sensitive film it is possible to take photographs in the dark of hot objects losing radiant heat.

Infrared radiation from hot objects is not itself warm. It does not warm the space through which it passes. But when infrared waves strike matter some are absorbed. They make the molecules of the absorbing material vibrate faster. That is to say, the material becomes hotter.

Rough black surfaces absorb more radiant heat, and emit more radiant heat, than do smooth shiny surfaces. Because of this oil storage tanks are painted silver, people wear white clothes for summer games, and solar panels are matt black.

Heat from the sun reaches us by radiation. Most of the space between the sun and Earth contains no matter, so there is nothing to absorb the infrared waves or other forms of radiation. But the sun's rays are absorbed by the atmosphere (see page 125), by the waters of Earth, by the ground, by all plants, by all animals and by other objects exposed to them — and so all things are warmed.

21.2 Temperature control

There are many situations in which temperature is regulated by controlling the movement of heat. To do this it is necessary to consider how heat travels.

Keeping things hot or cold

A vacuum flask is designed to control the movement of heat by all three methods.

1 The flask is made of glass and the stopper of plastic. Both are bad conductors of heat, so there is little movement of heat out of or into the flask by conduction.
2 The flask is double-walled, with a space between the walls (see Figure 21.2). All the air is removed from this space, forming a vacuum, then the glass is sealed. As there is no air in the space, there can be no air currents. So there can be no movement of heat out of or into the flask by convection.

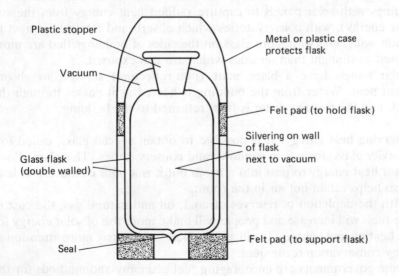

Plastic stopper

Metal or plastic case protects flask

Vacuum

Felt pad (to hold flask)

Silvering on wall of flask next to vacuum

Glass flask (double walled)

Seal

Felt pad (to support flask)

Figure 21.2 a vacuum flask. (Based on Keighley et al Mastering Physics**)**

3 The inner surface of the glass next to the vacuum is silvered. The glass, therefore, acts like a mirror and reflects radiant heat. Heat in the flask is kept in, and heat outside is kept out.

In these three ways the movement of heat out of and into the flask is controlled and the flask can be used to keep its contents either hot or cold.

Temperature control in buildings

To keep the temperature constant throughout a building you need:

1 A source of heat (such as a boiler or heater).
2 A means of distributing the heat (such as water-filled pipes and a water pump).
3 A means of dissipating heat in each room (such as radiators).
4 A thermostat to switch the boiler on and off (see Figure 4.5D).
5 Insulation to reduce heat loss from the building in cold weather and to reduce heat gain in hot weather.
6 Ventilators to expel excess heat, and in some countries refrigeration equipment to cool the air inside the building in very hot weather.

Heat energy sources. The fuels most commonly used as sources of heat energy are wood, coal, oil, gas and electricity. If electricity is used it is best to have a heater in each room so that a means of distributing heat is not needed and each appliance can be controlled separately by its own thermostat.

In an attempt to reduce the rate of depletion of reserves of coal, oil and natural gas, some governments are encouraging the construction of buildings with solar panels to capture radiant heat energy from the sun (solar energy), with conservatories which absorb and retain heat next to outside walls, and with more glass on the sides of buildings that are more exposed to sunlight than on sides which are less exposed.

Solar panels have a black matt (non-reflecting) surface, to absorb radiant heat. Water from the building is heated as it passes through the panel, and the warmer water is then returned to the building.

Conserving heat energy. It is possible to obtain special glass, called low emissivity glass, for use in windows and conservatories. This glass allows radiant heat energy to pass into a room but it is a poor conductor of heat and so helps retain hot air in the room.

With the depletion of reserves of coal, oil and natural gas, the cost of these fuels will increase and people will make more use of solar energy for both heating and lighting. It will also be necessary to pay more attention to energy conservation techniques.

Some governments are encouraging fuel economy and methods for the more efficient use of heat energy. For example, most electricity power stations are in rural areas and have large cooling towers to dissipate excess heat. If smaller power stations were constructed in urban areas, water could be cooled by passing it through community heating systems, to heat houses and other buildings. Similarly, in factories hot water produced in some chemical reactions can be passed through heat exchangers to heat raw materials or to heat buildings.

In the home, much heat is wasted. For example, with an open fire hot air carries the smoke up the chimney and much heat is lost from the building. Fires could be designed with efficient heat exchangers, so that more heat was retained in the home.

Other energy conservation measures include the use of thermostats, so that no room is heated above the required temperature, and the use of effective insulation techniques. Double glazing traps air between two sheets of glass and so helps to retain heat in a room (as do curtains and blinds). Still air, a poor conductor of heat, is trapped in most insulating materials used — for example, between the rafters in a roof space, to minimise heat loss through the roof.

Note that the sheets of glass used in double glazing should be 6 mm apart. With a wider gap convection currents in the air reduce the effectiveness of air as an insulating material. In contrast, when double glazing is used for sound insulation the sheets of glass should be further apart (12–25 mm).

Also note that the space in a cavity wall is not intended for heat insulation. It is for ventilation — to remove water vapour evaporating from the exposed bricks of the outer wall. This space is ventilated by an air current entering through air bricks just above the damp course (see Figure 4.7) and leaving at the top of the wall. This cavity does, however, help to keep the inside of a building cool in hot weather.

If people really wanted to conserve heat energy, and so reduce their heating costs, they would heat each room only when necessary and no more than necessary. They would make more use of clothing as a means of retaining body heat.

For all healthy people, in all except cold climates it is probably healthier to sleep in a well ventilated room, near an open window, than in a stuffy heated room with the window closed. It is best to rely on adequate bedding and bed clothes — including bed socks and a night cap if necessary — to keep you warm on cold nights.

Investigation 21.2   *Energy conservation*

Energy conservation is desirable to help reduce costs to the individual, the family, and the community, and to help conserve limited reserves of coal, oil and natural gas. One method of conserving energy is to switch off lights, and to work or play in daylight, whenever possible. Energy can also be conserved in ways suggested in this chapter. You may think of other ways.

> *You will need* only a notebook and a thermometer to begin your work.
>
> *Proceed as follows*
> Keep a record of the times when lights are on unnecessarily, and when the temperature is higher than you consider necessary in a building you use (for example, in your home, school or library). Make notes of what you think would be a suitable temperature for the kind of work done in that building or for each room if used for different purposes. For example, a low temperature might be desirable in a store room and in rooms in which people are active, and a higher temperature in a room used as an office in which people are not moving about very much. Suggest ways in which the use of energy could be reduced.

Temperature control in your body

The maintenance of a constantly high temperature throughout your body (see Figure 21.3) depends upon six things:

1 Whenever energy is transferred from one form to another there is some loss of heat energy. Heat is therefore produced, at all times, in all the cells of your body in their respiration (see page 168). It is produced especially in the cells of your liver and in your skeletal muscles when you are moving or shivering.
2 Heat is distributed throughout your body. This results from the presence of blood capillaries near to all cells and the rapid circulation of your blood (see page 220).
3 There are temperature receptors in your skin. Also, certain cells in your brain are sensitive to the temperature of the blood flowing through nearby capillaries. Nerves from your brain carry nerve impulses to the parts of your body, which make appropriate responses leading to increased heat production, heat conservation, or more rapid heat loss.
4 You lose heat each time you breathe out, and through your skin, especially in cold weather. Heat is also lost in faeces and urine. However, in cold weather the erector muscles of your hairs contract and, particularly in mammals with more hair than we have, more air is trapped between the hairs. This acts like a blanket and reduces heat loss. Also in cold weather, the arterioles in your skin are constricted and less blood flows through the blood capillaries near the surface (see Figure 9.7). The layer of fat deeper in the skin is then effective in reducing heat loss from the internal organs: your skin temperature is then lower than that of your body core.

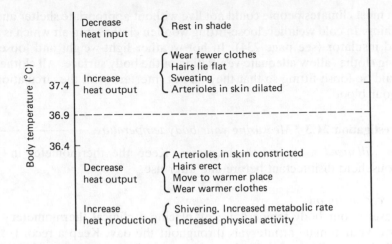

Figure 21.3 **physiological responses to variations in body temperature: at 0.5 °C above normal (dotted line), sweat is visible on the skin; and at 0.5 °C below normal, shivering starts. (From Barrass, R.** Human Biology Made Simple)

Note. Anyone suffering from shock (or cold) should not be given drinks containing alcohol because this drug causes dilation of arterioles in the skin and so both increases heat loss and reduces the blood supply to internal organs (including the brain).

5 In hot weather the hairs are not erect and so less air is trapped near the body surface. Also, more heat is carried to the skin because the arterioles in the skin are now dilated and more blood flows through the capillaries of the dermis (see Figure 9.7). The fat layer of the skin does not therefore act as an insulation layer in hot weather. These things facilitate heat loss, but there is also a cooling mechanism: sweating.

If your body temperature rises (even 0.2 to 0.5 °C) above normal body temperature (36.9 °C), sweat pours from the sweat glands and spreads over the surface of your skin. The evaporation of this sweat removes heat from your body. Note that water, compared with other substances, has a high latent heat. It takes large quantities of heat energy to change the state of water — in the melting of ice or the evaporation of water (see page 106).

6 We avoid extreme temperatures. Most mammals rest in the shade on hot sunny days. Some burrow in the ground where the temperature is less variable than at the surface. Some build nests. Some rest closer together when it is cold and so reduce the area of their exposed surface. Some migrate and others hibernate before the cold weather starts. Without appropriate behaviour temperature control would be impossible.

In most climates people could not live without appropriate shelter and clothing. In cold weather, loose-fitting woollen clothes trap air which is a good insulator (see page 234). In hot weather light-weight and loose-fitting clothes allow adequate ventilation of the body surface. All clothes should be loose-fitting so that they do not interfere with the circulation of your blood.

Investigation 21.3   *Measuring your body temperature*

You will need a clinical thermometer. Keep the thermometer in a household disinfectant before and after use.

Proceed as follows
Measure your body temperature by placing a clinical thermometer below your tongue at intervals throughout the day. Keep a record. 36.9 °C (98.4 °F) is described as our normal body temperature but you will find that your body temperature varies. Note things that seem to influence your temperature.

Homeothermy

Birds and mammals, the only animals that maintain a high and fairly contant body temperature, are called homeotherms (see Figure 21.3). All other animals are called poikilotherms: their body temperature is the same, or nearly the same, as the temperature of their environment.

There are advantages in homeothermy. For example, a homeotherm is always ready for action. Also, high temperature of the brain is considered to be necessary for the development of the memory. However, there are also disadvantages in homeothermy. Homeotherms need more food than poikilotherms because part of their food is used in heat production (heat energy is obtained from chemical energy). Also heat or cold kills most homeotherms if they cannot keep their body temperature near to their normal level.

In a hot and humid atmosphere people may suffer from overheating because they are unable to lose heat. This may cause *heat exhaustion* (fainting and even death). And in a hot dry atmosphere people may lose water (by sweating) until no further water loss is possible and sweating stops. Then over-heating causes *heat stroke* (collapse and even death).

In cold air your body temperature may fall. If you are inactive you start to shiver when the temperature of your internal organs (your body core) falls 0.5 °C below normal. Unless you move to a warmer place or put on warmer clothes your temperature may continue to fall. If your body core temperature falls to 35 °C (95 °F) or below you suffer from *hypothermia*. At first this causes uncontrollable shivering, then loss of coordination (see page 230), collapse and, unless properly treated, death. The symptoms of hypothermia are easily confused with drunkenness — especially if the person has been drinking and smells of alcohol — and death may then result from incorrect treatment.

Old people produce body heat slowly and are likely to suffer from hypothermia if they economise on food, clothing or heating. Sedatives may reduce their awareness of cold, and alcohol even increases the rate of heat loss (see page 239). If they are inactive, old people must rely on warm clothes, warm food and drinks, a suitable room temperature, and warm bedding.

For active, healthy people 18.3 °C (65 °F) is a comfortable sitting-room temperature. In a bedroom 12.8 °C (55 °F) is acceptable, but less than 10 °C (50 °F) is too low.

Babies, because they are so small, have a proportionately larger surface area than adults. Also, young babies cannot take exercise or move to another place. They can very quickly suffer from heat or cold, and should never be left with too many clothes and/or in a hot place; or with too few clothes and/or in a cold place. They should always be shaded from direct sunlight and properly clothed, according to the air temperature. A baby's cot should be in a ventilated room but not in a draught. And because of the risk of accidents the cot should not be placed below a window.

22

Producing more people

Because each individual has a limited lifespan, the continuance of any species depends on the production of new individuals of the same kind.

Sexual reproduction involves the formation and fusion of gametes (sex cells). In people there are two kinds of gametes: (1) small active cells called sperms, and (2) much larger immobile cells called eggs. Boys and men, who produce only sperms, are called males; and girls and women, who produce only eggs, are called females. Following the release of many sperms near an egg (see page 248), one sperm enters the egg and then the sperm nucleus fuses with the egg nucleus. This *fusion of gametes is called fertilisation*. The fertilised egg, called a zygote (Greek *zygosis* = joining) develops into a baby.

The nucleus of each body cell has twice as many chromosomes as the nucleus of a gamete. Human body cells (with 46 chromosomes each) are therefore called diploid cells (Greek *diplous* = double). Sperms and eggs (with 23 chromosomes each) are called haploid cells. The zygote, formed by the fusion of two haploid cells (one egg and one sperm), is therefore diploid (see Figure 22.1).

The type of cell division in which haploid cells are formed from diploid cells is called *meiosis* (Greek *meiosis* = a lessening). Contrast this with the division of other body cells (see mitosis, page 94) in which diploid cells are formed from diploid cells. Because meiosis in gamete formation is followed by the fusion of gametes: (1) the chromosome number for each species remains constant from generation to generation, and (2) half the chromosomes in each zygote come from the mother and half from the father.

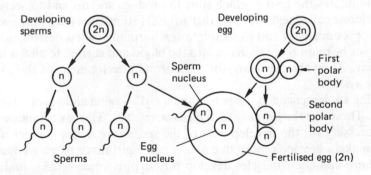

Figure 22.1 the results of meiosis in the testis and ovary, and the fusion of gametes forming a fertilised egg

22.1 The reproductive systems of men and women

A man has two testes, in the scrotal sac between the legs (see Figure 22.2). Very many sperms are produced in each testis, from about the age of 12. In sexual intercourse (see page 248) sperms pass along the sperm ducts and are mixed with the secretions of the prostate glands, seminal vesicles, and other glands, forming seminal fluid or semen. The urethra carries semen (and at other times urine, see page 215) to the outside. The tip of the penis is covered by a tubular fold of skin (the foreskin) unless this has been removed, usually soon after birth, in a minor operation called circumcision.

A woman has two ovaries, one on each side of the abdominal cavity (see Figure 22.2). From about the age of 11, one egg is released each month by one of the ovaries, into the funnel of the nearby oviduct. The egg moves down the oviduct and is likely to be fertilised if sperm are present. The uterus is a pear-shaped organ which projects into the vagina, the passage from the uterus to the outside. The projection of the uterus into the vagina is called the cervix. The outer and inner lips over the openings of the urethra and vagina, between the legs, are called the vulva. The clitoris, a small knob-like structure, is also between the inner lips of the vulva. A thin membrane, called the hymen, is across the opening of the vagina. The small hole in this membrane is enlarged if a girl uses internal tampons (to absorb blood lost in menstruation, see page 248) and the hymen is torn in sexual intercourse.

The parts of the reproductive system visible externally (the penis and scrotum of a man and the vulva of a woman) are called the genitalia or genitals. These parts should be washed every day.

Adolescence

The time when a boy is changing into a man, and a girl into a woman, at the start of adolescence, is called the age of puberty. The pituitary gland secretes a hormone called follicle stimulating hormone (FSH). In a boy, FSH stimulates the testes, which start to produce sperms and to secrete a sex hormone called testosterone. In a girl FSH stimulates the ovaries which start to produce eggs and to secrete a sex-hormone called oestrogen.

The sex hormones are secreted into the blood and carried to all tissues of the body. They are necessary for the normal development of the reproductive system.

During adolescence a boy's penis, and a girl's vagina and vulva, increase in size. These are the *primary sexual characteristics*. The sex hormones are also necessary for the development of the *secondary sexual characteristics* which make a boy look more like a man and a girl more like a woman. A boy's limb and chest muscles develop further, his voice breaks, and hair grows on his face and chest, in his armpits, near his genitals and in a triangle up to his navel. The shape of a girl's body changes as a result of

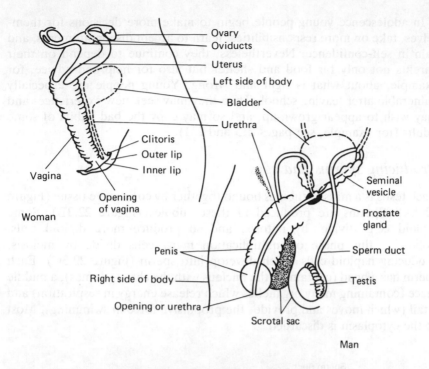

Figure 22.2 **reproductive system of a woman and a man. (From Barrass, R.**
Human Biology Made Simple)

deposition of fat in her breasts and hips. Hair grows in her armpits and around her genitals. The sex hormones also affect behaviour: boys take more interest in girls; and girls take more interest in boys.

Puberty is usually earlier in girls than in boys, and is at different ages in different races. A girl has her first period when she is 9–16 years old, so girls should be told about this monthly bleeding before they are nine. There is no equivalent event in the life of a boy, but the start of sperm production may be marked by a cream-coloured fluid (containing sperm) passing from his penis at night. These nocturnal emissions of seminal fluid happen from time to time, not regularly, and usually when he is dreaming of girls.

Puberty is marked by an increased rate of growth. No two people develop in quite the same way and the changes in puberty may be almost complete in some boys and girls at an age when the changes have hardly started in others. There is no need to worry, therefore, if you are an early or late developer. Differences in mental development are not closely related to differences in physical development. Late developers (mentally and physically) will be just as clever and just as physically developed at the end of adolescence as if they had developed early.

In adolescence young people begin to make more decisions for themselves, take on more responsibilities, learn to accept disappointments, and gain in self-confidence. Nevertheless, they continue to depend on their parents not only for food and shelter but also for help and advice, for example, about what is right and wrong. Young people are especially vulnerable after leaving school. Then they may seek new experiences and may wish to appear grown-up, and so may copy the bad habits of some adults (for example, see pages 215 and 231).

Producing sperms and eggs

Each testis is a mass of tubules bound together by connective tissue (Figure 22.3A). Sperms are produced in these tubules (Figure 22.3B). Some diploid cells divide, by mitosis, and so produce more diploid cells. Following this phase of multiplication many cells divide by meiosis, producing haploid cells which develop into sperm (Figure 22.3C). Each sperm has a head (containing the nucleus with the chromosomes), a middle piece (containing many organelles which release energy in respiration) and a tail (which moves and provides the propulsive force in swimming). Most of the cytoplasm is discarded.

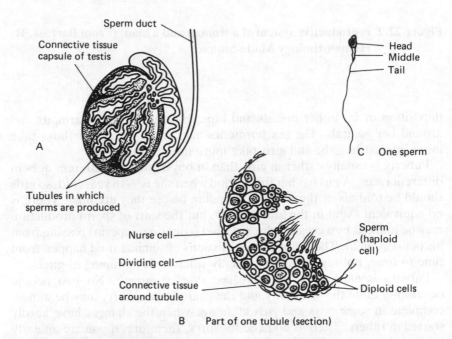

Figure 22.3 structure of testis of a mammal, and sperm production: (A) section of testis; (B) cross-section of one tubule; and (C) parts of one sperm. Diagram not to scale. (From Barrass, R. Human Biology Made Simple**)**

Eggs are produced in groups of diploid cells just below the surface of the ovary. When a girl is born she already has many of these groups of cells (Figure 22.4A) and no more are formed later. After puberty, each month some of these groups of cells develop further (Figure 22.4B). One cell covered by the others, grows and receives nutriment from the others, which are called follicle cells. The larger cell, a developing egg, is released when the follicle ruptures (Figure 22.4C) at about the middle of the 28-day cycle. This release of an egg is called ovulation. After this the follicle cells

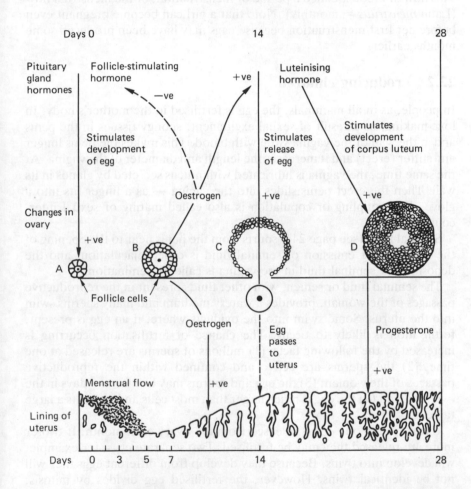

Figure 22.4 menstrual cycle: some changes in the pituitary gland, ovary and uterus. Effects of hormones are marked +ve (stimulates) or −ve (inhibits). At the end of each cycle a new cycle begins, except during pregnancy. A = group of cells; B = developing egg; C = release of egg; D = follicle cells form corpus luteum (yellow body). (From Barrass, R. Human Biology Made Simple)

continue to grow and divide, producing the corpus luteum (Figure 22.4D) which secretes another sex hormone (progesterone).

The cycle of activity in the ovary is controlled by hormones secreted by the pituitary gland, and hormones secreted by the ovary influence what is happening in both the pituitary and the uterus. These changes in the pituitary, ovary and uterus (and the mechanism of hormonal control) are shown in Figure 22.4. The uterus lining is discarded, with bleeding, after every 28-day cycle: that is to say, during the first few days of the next cycle. This flow of blood is called a period or menstruation, or the menstrual flow (Latin *menstruus* = monthly). Note that a girl can become pregnant even before her first menstruation because eggs may have been produced some months earlier.

22.2   Producing children

In people, as in all mammals, the egg is fertilised in the mother's body. In love-making, as a result of sexual excitement, spongy tissues in the penis and in the wall of the vagina swell with blood. This makes the penis longer and stiffer (erect) and it increases the length and diameter of the vagina. At the same time, the vagina is lubricated with mucus secreted by glands in its wall. Then the erect penis slides into the vagina — as a finger fits into a glove. This coupling or copulation is also called mating or sexual inter-course.

Seminal fluid (see page 244) spurts from the penis near to the opening of the uterus. This emission of seminal fluid is called ejaculation, and the deposition of seminal fluid in the vagina is called insemination.

The seminal fluid or semen, with other fluids present in the reproductive passages of the woman, provides a watery medium and many sperms swim into the uterus. Some swim into the oviducts where, if an egg is present, fertilisation is likely to occur. The chance of fertilisation occurring is increased by the following facts: (1) millions of sperms are released at one time, (2) the sperms are active and confined within the reproductive passages of the woman, (3) the egg and sperms may live for two days in the oviducts, and (4) the egg is much larger than most cells and presents a large target.

Usually, only one egg is released from the ovary each month. If two or more are released they may be fertilised. Two separate eggs, for example, will develop into twins. Because they develop from different eggs they will not be identical twins. However, the fertilised egg divides by mitosis, producing two cells which have identical sets of chromosomes (see page 95) and if these two cells separate they develop into separate but identical twins.

Usually, only one baby develops from one fertilised egg. The fertilised egg divides by mitosis and the cells produced keep on dividing until a ball of cells is formed. This becomes attached to the wall of the uterus — a process called implantation.

An embryo (developing individual) is produced from the ball of cells, as are the placenta, the umbilical cord and the membranes that enclose the embryo (Figure 22.5). The developing embryo is enclosed by these membranes, in a sac of fluid — like the embryos of all other animals the human embryo develops in a watery fluid. It is protected against water loss and mechanical shock by being in this fluid and by being carried within its mother.

Table 22.1   *The nine months of life before birth*

Time	Development and growth in uterus
0	Egg fertilised in oviduct. Development starts.
2 days	Fertilised egg has divided: two-cell stage.
5 days	Cell divisions continue. Embryo now in uterus.
8 days	Embryo of about 100 cells is implanted in wall of uterus
3 weeks	Heart, blood vessels and placenta have developed.
4 weeks	Head recognisable. Brain and spinal cord have developed.
5 weeks	Limbs and skeleton are developing.
6 weeks	Embryo, with a full set of organs, is now called a foetus.
2 months	Foetus about 25 mm long. Skeleton of cartilage but bone formation has started.
5 months	The mother can feel her baby moving its arms and legs.
7 months	Baby turns, head down, in uterus. Baby is complete and, if born prematurely, is likely to survive.
9 months	Baby is born about 266 days after fertilisation (= 280 days after its mother's last menstruation started).

The placenta is closely applied to the wall of the uterus. Here the blood of the mother is very close to the blood capillaries of the placenta (Figure 22.5). The umbilical cord is a lifeline through which an artery carries blood from the developing baby to the capillaries of the placenta and a vein carries blood back to the baby. In this blood, food materials and oxygen

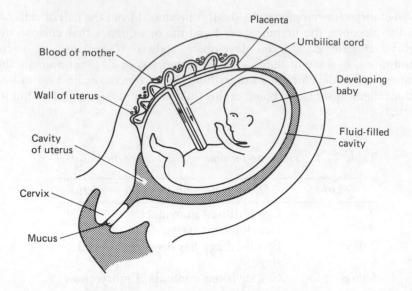

Figure 22.5 a baby developing in its mother's uterus. Parts of the mother are labelled on the left. (From Barrass, R. Human Biology Made Simple)

pass from the mother to the baby, and waste products (including urea and carbon dioxide) pass from the baby to the mother.

After about six weeks the embryo has a full set of organs and is called a foetus. The development of the human foetus from an embryo and its further growth before birth takes about nine months (see Table 22.1). This is called the gestation period (or pregnancy) and a woman who is carrying her baby in her uterus is described as pregnant. At the end of pregnancy the muscles in the wall of the uterus contract, bursting the membranes and releasing the fluid in which the baby has lived. Then the baby is forced through the distended cervix and vagina (Figure 22.6B). This process, called parturition or labour, is completed with the birth of the baby.

The baby starts to breathe immediately after birth although it is still connected to the placenta by the umbilical cord. The connection between the baby and the umbilical cord is ligatured and then cut (by a doctor or midwife), but the point of attachment remains (the navel: Figure 5.6). About 20 minutes after the baby is born, further contractions of the mother's uterus loosen the placenta, which comes away from the uterus (with the umbilical cord and membranes) as the afterbirth.

Breasts and breastfeeding

A girl's breasts start to develop in adolescence but milk is not produced by the mammary glands, in each breast, until the end of pregnancy. The

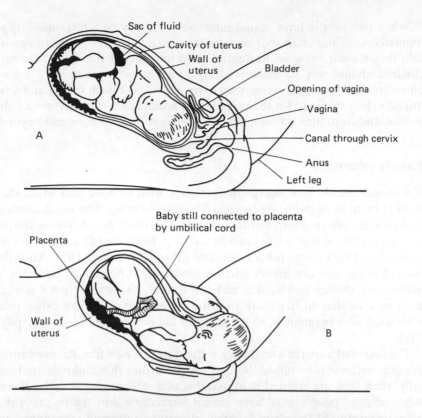

Sac of fluid
Cavity of uterus
Wall of uterus
Bladder
Opening of vagina
Vagina
Canal through cervix
Anus
Left leg
A

Baby still connected to placenta by umbilical cord
Placenta
Wall of uterus
B

Figure 22.6 two stages in the birth of a baby: (A) at the beginning of labour; and (B) baby emerging through the distended cervix and vagina. (From Barrass, R. Human Biology Made Simple)

secretion of milk starts soon after her baby is born, and breastfeeding should start as soon as possible (within six hours of birth).

Mother's milk is the natural food for a young baby — the best food it could have, containing not only all the essential nutrients, in the right proportions for a balanced diet (see page 199), but also antibodies produced by the mother, which help her baby resist infectious diseases. In addition, breastfeeding helps a mother to relax, and get her figure back after pregnancy, and brings the mother and her baby close together in a mutually satisfying and unique relationship.

Parental care

A man and woman may have sexual intercourse not only when they wish to have a baby but also as an expression of their love; and normally as part of a lasting relationship that enables them to care for their children.

When two people have sexual intercourse they have a responsibility to themselves, to the child that may be produced, and to others, to ensure that they do not bring an unwanted child into the world. That is to say, children should not be conceived by accident as a result of a casual encounter but only by parents who love and respect each other, and who intend to love and care for their children. Parental care starts before a baby is born and continues through childhood and adolescence — and beyond.

Family planning

If a couple decide how many children they wish to have and when they would like to have them, this is called family planning. The usual result is that families are smaller than they would otherwise be. A young couple may also prefer to wait a few years before they have their first child, to give themselves time to establish a home and get used to married life. Also, the risks of pregnancy are higher when a mother is in her teens: her baby is likely to be smaller and weaker and is less likely to survive than if it were born to a mother in her early twenties. Note that there are other risks associated with pregnancy, to the baby, if the parents are older (see page 259).

The fusion of a sperm with an egg is the start of a new life. An egg cannot develop unless it is fertilised. So one way to ensure that children are born only when they are wanted is to avoid sexual intercourse at other times. Alternatively, people may have sexual intercourse but try to prevent a baby being started (conceived). Such attempts to prevent conception are called methods of contraception (Latin *contra* = against).

Some methods of contraception

1 The man can cover his penis with a sheath of very thin rubber (a condom) which must be put on before there is any sexual contact.
2 The woman may insert a specially made cap or diaphragm into her vagina, before sexual intercourse, to cover the entrance to her uterus.
3 A woman may insert a spermicidal (sperm–killing) cream or tablet into her vagina before sexual intercourse, but these are best used with a sheath, cap or diaphragm.
4 A woman may take pills containing hormones that prevent the release of an egg from her ovaries. These pills can be obtained only from a doctor or clinic so that regular supervision and advice are available.
5 A doctor can insert a specially made flexible coil or loop of plastic into the woman's uterus. This prevents any fertilised eggs from developing, but can be removed if the woman decides she would like to have a baby.
6 The chances of having a baby are reduced if people avoid sexual intercourse near the middle of the woman's 28-day cycle, when an egg is most likely to be present in an oviduct (see Figure 22.4)

Note that withdrawal of the penis from the vagina before ejaculation is not a method of contraception because some sperms may pass from the penis before the bulk of the seminal fluid is ejaculated.

Abortion

A baby born about seven months after conception is likely to survive in spite of the premature birth; but it may have to be kept for a while in an incubator and given special care. If a developing baby is born too early for it to survive outside its mother's uterus, this is called a miscarriage or natural abortion.

It is also possible for a pregnancy to be ended by a doctor, especially if there is some risk to the mother's health or if the baby is known to have some inherited defect. Such an artificial termination of pregnancy is usually called simply an abortion. Abortion itself involves some risk to the mother, and it brings to an end the baby's life. Abortion is not therefore to be thought of as an alternative to contraception, as a means of family planning.

22.3 Sexual behaviour and disease

Diseases caught as a result of having sexual intercourse with an infected person are called sexually transmitted or venereal diseases (VD). You cannot tell if someone has one of these diseases. They may not know themselves because it takes some time for symptoms to develop.

Gonorrhoea is one of the most common diseases in the world. A few days after infection a man finds it painful to pass urine and he may see a discharge of yellowish matter from his penis. A woman may also have a yellowish discharge, which she is unlikely to see, but she may pass on the disease to others. Without treatment, gonorrhoea causes pain and may make a man or woman sterile (unable to have children). And a pregnant woman who has gonorrhoea may infect her child.

Syphilis causes sores on or near the genitals 10–90 days after infection. A man is likely to see these first symptoms, but a woman may not. Even without treatment the sores disappear, but months later skin rashes or mouth ulcers develop, followed years later by heart disease, blindness, madness or paralysis. At all stages the infection can be passed on to others. A pregnant woman may infect her baby, which may be born dead or blind or with a damaged heart.

If you think you may have caught gonorrhoea or syphilis, see a doctor immediately, preferably before any symptoms develop. These diseases, caused by bacteria, can be treated using antibiotics.

AIDS, a virus disease caught during close bodily contact with an infected person of the same or opposite sex, or by using a hypodermic needle contaminated with infected blood, destroys the body's defences against other infectious diseases. Symptoms of AIDS do not develop at once, so an infected person may not know that he or she has the disease. But it seems likely that anyone infected with the AIDS virus will die, possibly within ten years of infection, because as yet there is no cure.

Use of contraceptive pills has removed the fear of pregnancy that used to deter many people from sexual intercourse outside marriage. Other factors contributing to increased promiscuity include the earlier independence of many young people and the increased consumption of alcohol and other drugs that reduce self-control. A sheath (condom) gives some protection against sexually transmitted diseases; but the best way to avoid infection is to have sexual intercourse only as part of a lasting relationship, as in marriage.

23 Genetics and evolution

There is great variation in each species. You will have observed that no two people, even identical twins, are exactly alike.

Some differences between individuals are the result of external or environmental influences during development and growth. For example, a child's physical and mental development may be adversely affected by undernourishment (see page 201), by a deficiency of iodine in its diet (see Table 18.1) or by its mother smoking or taking drugs — such as alcohol — when the child was developing in her uterus.

Many differences and similarities between individuals are due not to environmental influences but to heredity. All people resemble other people in most respects, and some children strongly resemble one of their parents, or one of their grandparents. There may be particular features that are recognised as family characteristics.

The egg and the sperm are the only link between one generation and the next. Therefore, those characteristics of the parents that can be recognised in later generations must be represented in some way in the gametes. The study of the mechanism of heredity is called genetics (Greek *genesis* = descent).

23.1 Genetics: the study of heredity

The modern science of genetics began with the work of an Austrian monk, Gregor Mendel, who crossed pure-breeding varieties of the garden pea *Pisum sativum*. Other scientists have carried out similar experiments with many kinds of plants and animals and we now know that the mechanism of inheritance is similar in them all. We also know from medical records, and from studies of several generations of many families, that inheritance in people occurs in a similar way to inheritance in other animals.

Hybridisation: crossing pure-bred varieties

If a pure-bred black mouse is mated with a pure-bred brown mouse (labelled parents in Figure 23.1) all the offspring (labelled offspring 1 in Figure 23.1) are black. If two of these offspring 1 mice are mated, some of their young (labelled offspring 2 in Figure 23.1) are black and some are brown. There are about three times as many black mice as brown mice in this generation.

The crossing of pure-bred varieties is called hybridisation and the offspring which result from such a cross are called hybrids. A cross such as this, involving one pair of contrasted characteristics (black and brown) is called a monohybrid cross (Greek *monos* = single; Latin *hybrida* = cross).

Because the gametes are the only link between one generation and the next, the gametes of the mouse that is pure-bred for black coat colour must all contain a factor for black (represented by the capital B in Figure 23.1). Similarly, the gametes of a mouse that is pure-bred for brown coat colour must all contain a factor for brown (b in Figure 23.1). These factors are now called genes.

Because the offspring 1 mice are formed following a cross between pure-bred parents, after fusion of gametes containing the gene for black (B) with gametes containing the gene for brown (b), these offspring must have both genes (Bb).

A capital B is used as the symbol for the gene for black and a small b as the symbol for the gene for brown, because when B and b are present

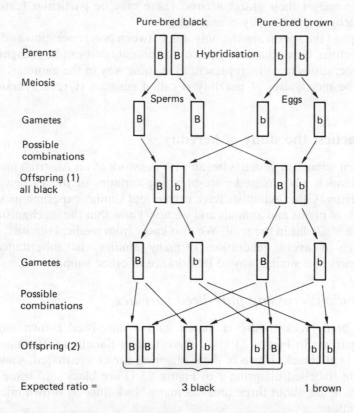

Figure 23.1 hybridisation: crossing pure-bred varieties of mice. One gene is represented on each chromosome (B for black, b for brown)

together (Bb in Figure 23.1) the mouse is black. B masks the effect of b and we say, therefore, that the gene for black is dominant to that for brown — which is said to be recessive.

If some of the sperms of the offspring 1 males have a gene for black (B) and others a gene for brown (b), and if the same applies to the eggs produced by the females of this generation, then there are four possible combinations of these sperms and eggs: BB, Bb, bB and bb.

If the sperms and eggs fuse at random, and each of these combinations is equally likely, we should expect there to be about three times as many black mice as brown mice in the next generation (offspring 2 in Figure 23.1).

A test cross. To find out if this interpretation is correct we need a method of testing the genetic constitution of the black mice (offspring 2 black mice in Figure 23.1). Such a test cross, which is also called a back cross because the black mouse with the unknown genetic constitution is crossed with a pure-bred brown mouse like that of the parental generation, is represented in Figure 23.2

If the black mouse has the genes BB all its offspring will be black. But if it has the genes Bb about half its offspring will be black and the rest brown. Using this test, the interpretation shown in Figure 23.1 can be proved to be correct. That is to say, about one third of the offspring 2 black mice have the genetic constitution BB and about two thirds Bb.

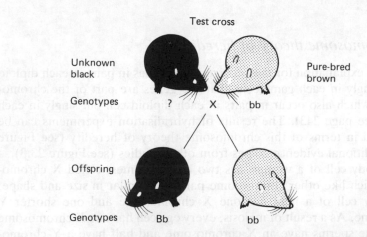

Figure 23.2 a test cross, sometimes called a back cross. The animal to be tested is crossed with a double recessive from the original parental cross. If some of the offspring are brown, the genotype of the unknown animal must be Bb. If both parents were homozygous for coat colour all the offspring would be black. (From Barrass, R. Modern Biology Made Simple)

Note. Genes are present in pairs in the body cells of each generation (BB, Bb or bb) but gametes have either the gene for black (B) or the gene for brown (b). Because these are alternatives they are called alleles.

The following law is based on Mendel's experiments on plant hybridisation, but from the experiment with mice you may reach the same conclusion: *of a pair of contrasted characteristics only one can be represented by its allele in any gamete*. This is known as Mendel's *law of segregation of alleles*.

We can distinguish between the genetic constitution of an organism (the set of genes present in the fertilised egg and in each body cell) and the outward appearance of the organism. For example, black mice have the allele combinations for coat colour BB or Bb even though they look alike. Clearly, organisms with different combinations of alleles (different genotypes) may develop into similar phenotypes (Greek *phainein* = to appear).

The alleles that occur in a pair may be alike (BB or bb) when the genotype is described as homozygous for this characteristic (Greek *homos* = alike; *zygosis* = joining) or they may be different (Bb) when the genotype is described as heterozygous for this characteristic (Greek *hetero* = different).

In the heterozygote a recessive allele is masked by a dominant allele, but note that the recessive allele must remain distinct in the heterozygote because its effects appear in later generations (see Figure 23.1). The test cross enables us to distinguish individuals that are homozygous from those that are heterozygous with respect to a particular characteristic.

The chromosome theory of heredity

A possible explanation for the occurrence of genes in pairs in each diploid cell but singly in each gamete, is that the genes are part of the chromosomes — which also occur in pairs in each diploid cell but singly in each gamete (see page 243). The results of hybridisation experiments can be understood in terms of this chromosome theory of heredity (see Figure 23.1). Additional evidence comes from other studies (see Figure 23.3).

Each body cell of a woman has two chromosomes, called X chromosomes, which like other chromosome pairs are similar in size and shape. Each body cell of a man has one X chromosome and one shorter Y chromosome. As a result of meiosis, every egg cell has one X chromosome but half the sperms have an X chromosome and half have a Y chromosome. An egg (with an X chromosome), therefore, is as likely to be fertilised by a sperm with an X chromosome (forming a zygote with two X chromosomes which develops into a girl) as by a sperm with a Y chromosome (forming a zygote with one X and one Y chromosome which develops into a boy).

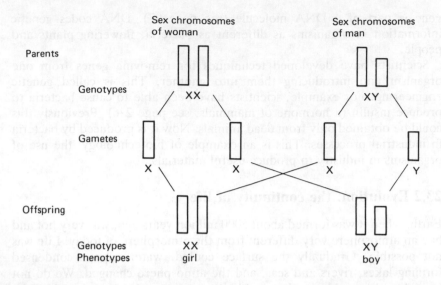

Figure 23.3 Sex determination in people: XX = female, XY = male. (From Barrass, R. Modern Biology Made Simple)

Genetic counselling

Most children have only two sex chromosomes in each body cell. But as a result of an unequal sharing of chromosomes in meiosis a few boys have an extra X or an extra Y chromosome (XXY or XYY), and a few girls have an extra X chromosome (XXX). An extra X chromosome results in a slight decrease in intelligence, and an extra Y chromosome results in a slight increase in aggressiveness.

The chromosome pairs are numbered 1 to 22 (plus XX or XY), making a total of 46 in a diploid cell. However, some individuals have an extra chromosome 21. Such children can be recognised by their facial characteristics, they may suffer from heart defects and all are mentally retarded. This condition, first described by an English physician called Down in 1866 is known as Down's syndrome. The risk of a child being born with this condition increases with the age of the mother (and father).

With increasing knowledge of the mechanism of heredity, it is possible to obtain information and advice on inherited conditions from doctors who specialise in genetic counselling.

Genetic engineering

It is clear from such evidence that the chromosomes do carry genetic information from generation to generation, and we now know that each gene is part of a DNA molecule (see page 97). DNA codes genetic information in organisms as different as bacteria, flowering plants and people.

Scientists have developed techniques for removing genes from one organism and introducing them into another. This is called genetic engineering. For example, scientists have been able to cause bacteria to produce insulin (a hormone of mammals, see page 226). Previously this could be obtained only from dead animals. Now it is produced by bacteria in industrial processes. This is an example of biotechnology: the use of organisms in industry to produce useful materials.

23.2 Evolution: the continuity of life

Earth, when it was formed about 5000 million years ago, was very hot and had an atmosphere very different from the atmosphere of today. Life was not possible. Gradually the surface cooled, water vapour condensed forming lakes, rivers and seas, and the atmosphere changed. We do not know how living things originated but there were microbes, thought to be similar to the bacteria of today, living on Earth about 3750 million years ago.

Evidence from scientific studies that the species living today are different from those that lived in the past was summarised by Charles Darwin in his book *The Origin of Species* published in 1859.

Evidence from the study of fossils

The only remaining direct evidence of life in past ages comes from the study of shells, bones and other hard materials preserved in sedimentary rocks. These remains, along with casts of organisms, and such things as footprints preserved as impressions, are all called fossils. And the fossils in rocks of different ages are a record of life on Earth (see Figure 23.4).

A logical interpretation of the fossil record, which is in accordance with all other evidence, is that organisms which are present later in the fossil record are the descendants of pre-existing organisms. For example, the first animals with backbones, which are now extinct, were fishes preserved in rocks formed about 450 million years ago. The first amphibians, the first reptiles and the first birds and mammals, were preserved in rocks formed more and more recently.

The fishes that lived so long ago could be the ancestors of all present-day fishes, and also of the first amphibians. The first amphibians could be the ancestors not only of the amphibia of today but also of the reptiles. And the first reptiles could be the ancestors not only of the later reptiles but also of birds and mammals.

Most organisms decay after death, or are consumed. As a result, very few are preserved. The fossil record, therefore, is incomplete. Darwin compared it to a book from which many pages are missing and on those

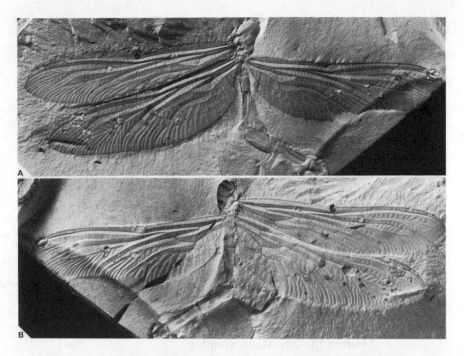

Figure 23.4 a fossil hawker dragonfly *Valdaeshna* **found in a frost-shattered rock by Dr E. A. Jarzembowski. This insect lived about 125 million years ago. Its wingspan is 85 mm. Reproduced from Palaeontology 31 (3) with the permission of the editor and of Dr Jarzembowski**

that remain only a few words are legible. However, even from the incomplete evidence provided by fossils we can conclude that some species that lived in the past, which we know only as fossils, have become extinct. Also, it seems clear that some present-day species did not exist in the past. This is why Darwin called his book *The Origin of Species*.

Evidence from comparative anatomy

The bones of the forelimbs of all four-footed animals with backbones are arranged in a similar pattern (see Figure 23.5). This is what you might expect if all the reptiles, birds and mammals, as well as all the amphibia living today, are descended from the first amphibia. That is to say, such anatomical evidence from the study of living animals with backbones can be interpreted in the same way as the fossil evidence.

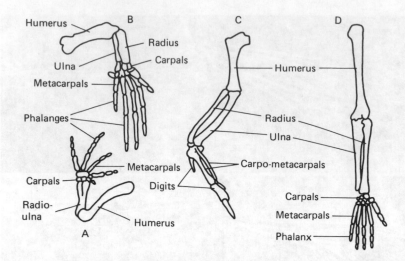

Figure 23.5 bones of the forelimbs of four land-living vertebrates (animals with backbones): (A) leg of frog *Rana*; (B) flipper of turtle *Chelone*: (C) wing of bird *Columba*; and (D) arm of a man or woman *Homo*. Drawings not to same scale. (From Barrass, R. Modern Biology Made Simple)

Evidence from the study of classification

The natural system of classification devised by Linnaeus (see page 21) can be understood in terms of the history of any group of organisms. A logical explanation for the fact that organisms can be placed in groups, according to the things they have in common which distinguish them from the members of other groups, is that all members of any group have a common ancestry. Similarities between the species in a genus are to be expected if they are all descended from the same ancestral stock. Similarities between the genera in a family are also to be expected, even if they are more distantly related than the species in a genus. And the same argument applies to the families in an order, to the orders in a class, to the classes in a phylum, and to the phyla in each kingdom.

Evidence from the study of geographical distribution

There are different species of the genus *Canis* in different parts of the word:

Canis lupus	wolf	Northern temperate regions
Canis latrans	coyote	North and Central America
Canis dingo	dingo	Australia
Canis aureus	golden jackal	Africa and Middle East
Canis familiaris	domestic dog	worldwide

A possible explanation for the occurrence of different species of dogs in different parts of the world, and for the occurrence of different races of people in different parts of the world, is that when members of a species are isolated in geographically separate groups, the members of each group may change in appearance and physiology over many generations. If this is so, then the longer they are isolated the more different they might be expected to become.

Evidence from the work of plant and animal breeders

All crop plants, farm animals and domestic animals are descended from wild species. They have been changed over thousands of years, and are still being changed, as a result of people selecting individuals that best serve some purpose and using these to produce the next generation.

As a result of the use of our knowledge of genetics in plant and animal breeding, since 1900, much higher yielding varieties of plants and animals are available to farmers.

Plant and animal breeding is a continuing process of selection that has resulted in the modification of species. That is to say, it has resulted in organic evolution.

Evidence from cytology, biochemistry and genetics

Further observations made since Darwin's time, in the new science subjects cytology (study of cells), biochemistry and genetics, can also be interpreted as evidence of organic evolution.

Similarities in the structure of cells from organisms as different as flowering plants, animals and protists, are to be expected if all these organisms are descendants of a common ancestor.

The similarities in the metabolic pathways, for example those involved in respiration, of organisms classified in different kingdoms, are also to be expected if the organisms are descendants of a common ancestor.

The mechanism of inheritance is similar in organisms as different as pea plants and mice. In both, information is coded in DNA molecules. This too is to be expected if these very different organisms have a common ancestor — that lived more than 1000 million years ago.

The mechanism of evolution

In 1858, a year before the publication of *The Origin of Species*, Charles Darwin and Alfred Russel Wallace suggested that as species can be changed by plant and animal breeders — by artificial selection — it is possible that species change in nature — by a process of natural selection. They suggested that the mechanism of natural selection might be as follows.

1 Every species has a great potential for increase. That is to say, each female may have many young.

2 In each species, numbers vary from year to year but there is not the constant increase in numbers that would be expected if all young animals survived and reproduced.
3 There is great variation in every natural population of organisms.
4 Most offspring do not survive to maturity. Therefore they do not reproduce. Those that do reproduce are likely to include many of those that are best adapted or best fitted to survive.

In other words, Darwin and Wallace suggested that there was a natural selection which resulted in the survival of the best adapted individuals in each generation.

In this changing world, natural selection could account for: (1) the phenomenon of adaptation to mode of life, (2) organic evolution — the gradual modification of species, and (3) the extinction of organisms that are no longer adapted for survival.

Darwin did not suggest that people were descended from apes or that any present-day species was descended from any other present-day species. He did suggest, however, that apes and people might be descended from a common ancestor. This would acount for the similarities between them.

23.3 What is life?

There is no simple definition of life, but we can list some features or characteristics which distinguish living from non–living things.
1 Each organism passes through phases of development, growth and maturity, and has a characteristic shape at each phase of its life.
2 Organisms are of different kinds, called species.
3 The shape of an organism is influenced by the environment in which it lives and this is one reason for variation within each species.
4 The shape and mode of life of each organism is appropriate to the place in which it lives — we say it is adapted to its environment.
5 Similar species can be arranged in groups, in a natural classification.
6 Organisms are either autotrophic (produce organic molecules) or heterotrophic (use organic molecules produced by other organisms).
7 Organisms are distinct from their environment but there is a constant exchange of energy and matter between each organism and its environment and some absorbed materials are assimilated (become part of the organism).
8 All organisms obtain energy from molecules by respiration; and excrete waste products of their metabolism.
9 Organisms are sensitive to stimuli in their external and internal environments, and they make appropriate responses to many of these stimuli and so increase their chances of survival.
10 Organisms are either single cells or are composed of many cells plus extracellular materials produced by the cells.

11 The parts of a cell and the parts of each organism are interdependent and, working together, they maintain the constant composition of each cell and of the organism as a whole — they maintain homeostasis (see page 230).

12 Organisms do not live alone. They interact with other members of their own species and with members of other species.

13 Each organism has a limited life span but reproduction and dispersal make possible the maintenance of the species.

14 Organisms are produced only from pre-existing organisms. The similarities between members of a species, and many of the differences, are the result of the passage of information (coded in nucleic acid molecules, see page 95) from generation to generation (see page 255).

15 Adaptation is a condition for survival. If any individual is not adapted it dies. If none of the individuals of a species is adapted the species becomes extinct.

16 Species that continue to be adapted in a changing environment may themselves change. This process of change in organisms is called organic evolution.

17 Life exists on Earth but, as far as we know, nowhere else.

24 Earth's climate and the weather

The surface rocks of Earth are covered in some places by soil, and in others by the waters of lakes, rivers, seas and oceans. The land, including lakes and rivers, covers about one third of Earth. Sea water covers two thirds.

Air, Earth's atmosphere, is a mixture of gases held in place by Earth's gravitational pull (see page 115). About 99 per cent of the air is within 40 km of Earth's surface.

Investigation 24.1 *Prepare a diagram to scale, to represent the diameter of Earth and the height of the atmosphere*

You will need unlined paper, a sharp pencil and a drawing compass.

Proceed as follows
1 Work to a scale of 1 cm = 1000 km.
2 Draw a circle to represent Earth (radius about 6367 km).
3 Draw another circle, with the same centre, to represent the upper limit of 99 per cent of Earth's atmosphere (about 40 km above sea level). You will not find this easy, because the second circle will be only 0.4 mm outside the first.

All life on Earth depends on this comparatively thin layer of air, and the relatively constant composition of air (see Figure 11.1) depends on the activities of living organisms (see page 125).

The air, soil and water which envelop Earth, and in which organisms live, is called the *biosphere*. As far as we know, life exists nowhere else in the universe.

24.1 Climate

Temperature

Earth is almost spherical (diameter at equator = 12 756 km and the axis through the poles = 12 713 km). As a result, identical amounts of radiant energy from the sun are spread over a much greater area at the

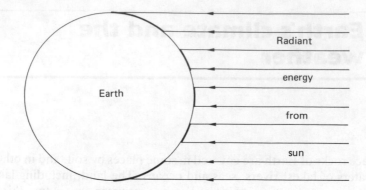

Figure 24.1 identical amounts of radiant energy from the sun are spread over a much larger area at the poles than at the equator

poles than at the tropics (see Figure 24.1). So the polar regions are heated least and the tropics most. This is one major influence on climate.

Investigation 24.2 *Areas illuminated by a light source*

You will need a lamp, a piece of card with a 2.5 cm diameter hole cut in the centre, and graph paper marked in millimetre squares.

Proceed as follows
1 Fix the card vertically, in front of the lamp, in a darkened room.
2 Switch on the lamp and hold the graph paper in the beam, first vertically and then at the same distance from the lamp but at an angle of 45° to the light. Compare the illuminated areas.

Because Earth's axis is at an angle to the sun, day length varies with the season much more at the poles than at the equator (see page 12). This accounts for the seasonal differences in climate which make polar and temperate regions much colder in winter than in summer.

These differences in exposure to the sun's radiant energy, resulting in great differences in the temperature of different regions of Earth's surface, and at different times of each year, are recognised when we describe parts of Earth as tropical, subtropical, temperate, and polar.

The influence of climate (especially temperature and rainfall) is indicated in the names we give to world regions. For example, some regions are described as equatorial forest, subtropical grassland (savanna), and desert. Depending on its climate, each region has a characteristic soil and vegetation type. And with each type of vegetation you associate certain kinds of animal. That is to say, each region has a characteristic flora and fauna, so each region can be thought of as a biome (where kinds of plant and animal characteristic of the region, live together).

Differences in temperature in different parts of Earth, which influence the development of vegetation and soils, also result from such things as altitude, aspect, and nearby ocean currents.

Winds and rain

The other major influence on the climate, vegetation and animal life of any region is rainfall. This depends partly on the direction of the prevailing winds — humid on-shore winds bring rain to inland areas whereas winds blowing from the centres of continental land masses are dry.

Earth's atmosphere extends hundreds of kilometres above sea level, and is in layers marked by differences in temperature. The lowest layer, in which the changes we call weather occur, is called the troposphere. This extends for about 12 km above sea level and contains most of the mass of air.

The study of Earth's atmosphere is called meteorology. It is of great interest because of the value of meteorological data in weather forecasting.

Winds. Differences in the temperature of latitudinal zones, affecting the temperature of the air by conduction, account for: (1) the differences in the density of air which cause the air currents we feel as winds, and (2) the existence of separate air masses in different latitudes and over the oceans and continents, described as polar maritime, tropical maritime, polar continental, and tropical continental.

As you would expect, air heated near the equator expands and so is less dense. It therefore rises, and cooler air moves in from both north and south towards the equator. Because of Earth's rotation (towards the east), the direction of these winds is towards the west (see Figure 24.2).

Cold air descends at the poles and there is a circulation of air in tropical and polar regions. The air circulating in tropical and polar regions results in a circulation of air in temperate latitudes. The circulation of air in the troposphere is shown in Figure 24.2 in relation to regions of high and low air pressure.

This simple explanation of wind direction near Earth's surface is complicated by the existence of large areas of land (continents) and large areas of sea (oceans). In summer, in north temperate latitudes, the land absorbs radiant heat from the sun. The air above is heated by conduction, resulting in its expansion. This creates an area of low atmospheric pressure (see Figure 24.3A). The less dense air rises, and air is drawn in towards the centre of the land mass. On-shore winds are represented by arrows in Figure 24.3A). In winter the land surface cools faster than the oceans, having opposite effects to those described above (see Figure 24.3B). This is why coastal areas are generally warmer in winter and cooler in summer, compared with further inland.

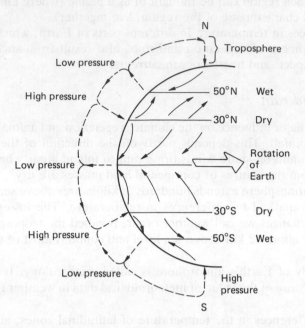

Figure 24.2 air currents in the troposphere. (Based on Hulbert, J. All About Weather W. H. Allen, London)

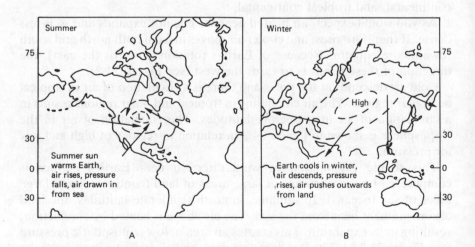

Figure 24.3 on-shore and off-shore winds. (Based on Hulbert, J. All About Weather **W. H. Allen, London)**

A more local effect in coastal areas on hot summer days with clear blue skies, is the rapid heating of the land by day and its rapid cooling after sunset. This may result in on-shore and off-shore breezes (see Figure 24.4) in the daytime and night-time respectively.

The effects shown in Figure 24.3 and Figure 24.4 are due to the fact that water has a very large specific heat capacity compared with other substances (see page 108).

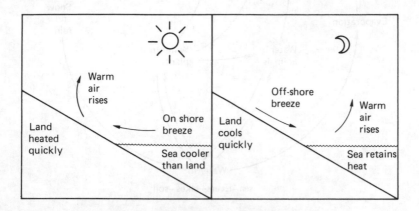

Figure 24.4 daytime sea breeze; night-time land breeze

The water cycle. Water covers about three quarters of Earth's surface and it seeps into the ground. In most places on land, therefore, it is possible to bore a hole and find water. The level at which this water occurs in the ground is called the water table.

Water evaporates from all water surfaces and is present in air as water vapour (an invisible gas, see page 115). Warm moist air rises and when it cools the water molecules come together (condense), forming water droplets. A mist is formed when this occurs at ground level, and a cloud when it occurs high above the ground.

Water droplets in the clouds combine to form larger drops which fall as rain, snow or hail — returning water to the soil, rivers, lakes and seas. This circulation of water in nature is called the water cycle (see Figure 24.5). Although most of the water vapour in the atmosphere comes from the sea, rain is not salty. This is because evaporation and condensation purifies the water (as in distillation, see page 85).

24.2 Weather forecasting

On a weather map, places of equal atmospheric pressure (see page 121) are joined by lines called isobars (see Figure 24.6) just as on a relief map places of equal height are joined by contour lines. The regions of highest pressure

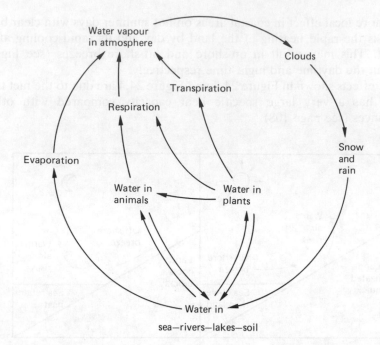

Figure 24.5 the water cycle. (From Barrass, R. Human Biology Made Simple)

are marked high, and those of lowest pressure are marked low. Regions of high pressure are called anticyclones. Gentle winds circulate around the high pressure centre (anticlockwise in the southern hemisphere; clockwise in the northern hemisphere). Days are dry and sunny. Regions of low pressure are called cyclones or depressions. Warm humid air rising in the centre results in strong winds, swirling towards the centre (clockwise in the southern hemisphere, anticlockwise in the northern hemisphere). Days are cloudy and there may be driving rain, hail or snow.

The isobars are usually closer together around low than around high pressure regions. The closer they are, the larger the differences in atmospheric pressure over short distances — so the stronger the winds.

Cold dense air in the temperate latitudes does not easily mix with warmer less dense air. As a result, air masses (see page 269) move but remain distinct (see Figure 24.6 and Figure 24.7). Relatively narrow transition zones where cold air masses replace warm air masses are called cold fronts. These may be associated with short periods of heavy rain, hail or snow. Wider transition zones are formed when warm air masses replace cold air masses. These are called warm fronts, and may be associated with prolonged light rain.

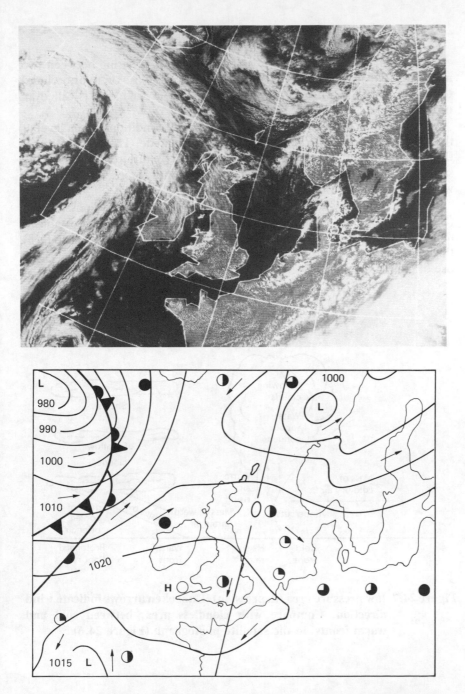

Figure 24.6 satellite photograph of part of the North Atlantic and N.W. Europe; and for comparison a barometric chart of the same area at the same time (August). L = low pressure; H = high pressure; arrows indicate wind direction; ▼▼ = cold front; ⬤⬤ = warm front; ▼⬤ = occluded front; ● = complete cloud cover; ◐ = 50% cloud over; ○ = no cloud. (Photograph courtesy of University of Dundee, Scotland; and barometric chart based on European Meteorological Bulletin)

Note in Figure 24.7 that as a result of the warm air rising between two masses of cold air, the cold front will catch up with (close on to or occlude) the warm front. The result, as two masses of cold air meet, is called an occluded front.

Weather maps provide a record of atmospheric conditions at sea level at the time the meteorological measurements were made. The preparation of these maps depends on observations being made in different places at the same time, and the rapid communication of the data recorded to the places where the maps are prepared.

The preparation of weather maps, therefore, has been made possible by: (1) the development of suitable measuring instruments, (2) the establishment of a network of meteorological stations where data are recorded, (3) the development of methods of rapid communication, and recently (4) the use of computers to analyse the data and prepare maps quickly.

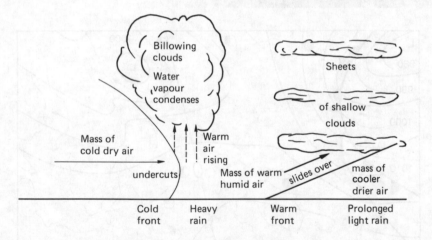

Figure 24.7 low pressure area (depression). Unbroken arrows indicate wind direction. Compare with cloudless area, between cold and warm fronts, in the satellite photograph (Figure 24.6)

Weather maps must be prepared quickly, because weather forecasting involves predicting, from recent observations, how conditions are likely to change in the next few hours or perhaps in the next few days. In these respects, meteorological charts differ from most other maps — which remain useful for years.

24.3 Above the troposphere

The layer of Earth's atmosphere above the troposphere is the stratosphere. Here, and in higher layers, short-wave electromagnetic radiation from the sun ionises the rarified gases. In 1901 an Italian scientist, Guglielmo Marconi, discovered that radio waves transmitted from Earth were reflected from the sky. They are reflected by a layer of ionised gases, called the ionosphere, which extends from about 80 km to about 300 km from Earth's surface. One of the ionised gases, formed from oxygen, is ozone. The stratosphere, therefore, is also called the ozone layer.

Destruction of the ozone layer

Ozone absorbs ultraviolet rays and reduces the amount of harmful radiation reaching the biosphere. Scientists have become increasingly concerned about the destruction of this protective ozone layer by man-made chemicals. The layer is thinnest over Earth's poles, and is thinner in winter than in summer. Holes in the ozone layer were first reported by Joe Farman of the British Antarctic Survey in 1985. The increase in ultraviolet radiation reaching Earth is expected to cause an increase in skin cancer, especially in white-skinned people.

The destruction of the ozone layer is due to compounds of chlorine, fluorine and carbon called chlorofluorocarbons (or CFCs) which are used in aerosols and as coolants in refrigerators. CFCs are very stable molecules. They rise unchanged in the atmosphere, and there combine with and so destroy ozone.

The greenhouse effect

Earth is warmed by the sun, then the atmosphere is warmed by conduction. The atmosphere also acts like a blanket — or like the glass of a greenhouse, absorbing infrared radiation from Earth (see page 234) and so retaining heat near Earth. This greenhouse effect is not new, and is largely due to the presence of water vapour and carbon dioxide in air. Only 0.03 per cent of the atmosphere is carbon dioxide (see Figure 11.1) but without this gas the air temperature would be over 30 °C lower than it is. But, probably as a result of burning fossil fuels, and cutting down trees in equatorial forests that previously removed much carbon dioxide from air in their photosynthesis, the amount of carbon dioxide in air has increased since the industrial revolution. Other air pollutants (for example, methane and nitrous oxide) are probably enhancing this greenhouse effect and there is much concern about global warming which may result. For example, if the polar ice caps melt there will be a gradual increase in sea level which could flood many cities and fertile lowland areas.

Earthquakes are localised movements of Earth's surface rocks. They cause vibrations that may do great damage to buildings, cause landslides, and result in great loss of life. When an earthquake occurs, vibrations pass through the whole planet and also pass round the circumference in the surface rocks (the crust, see Figure 25.1). The study of these shock waves, or seismic waves, is called seismology.

There are two kinds of shock wave. Primary waves are longitudinal waves (like sound waves; also see page 178). These travel in straight lines, but are refracted when they pass from solid to liquid or vice-versa. Secondary waves are transverse waves (like electromagnetic waves, see page 177) which also travel in straight lines through solid material, but cannot pass through liquids.

From studies of these two types of wave, radiating from the central point of an earthquake, scientists have concluded that: (1) Earth's surface rocks are a thin crust on (2) the denser material of a thicker semi-liquid mantle, and that (3) there is an even denser metallic core (mainly of iron) which may be entirely liquid or made up of outer liquid and inner solid parts (see Figure 25.1).

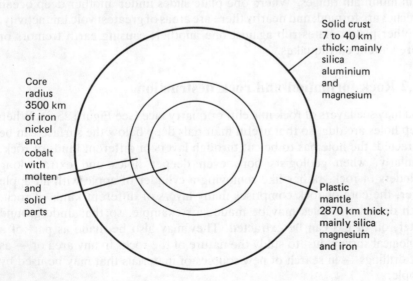

Solid crust 7 to 40 km thick; mainly silica aluminium and magnesium

Core radius 3500 km of iron nickel and cobalt with molten and solid parts

Plastic mantle 2870 km thick; mainly silica magnesium and iron

Figure 25.1 Earth's crust, mantle and core

25.1 Earth movements

The first maps of the whole world were prepared in the sixteenth century. In about 1620, Francis Bacon, an English philosopher, pointed out that the shapes of the continents were like pieces of a jigsaw puzzle. If cut out, from a map, they could be fitted together — like one land mass.

In 1912, Alfred Wegener, a German, brought together evidence in support of the hypothesis that the continents were once part of a single land mass. This evidence came from different scientific studies but, even as late as 1960, most geophysicists did not accept that this was anything more than an interesting hypothesis.

However, further evidence, collected in the 1960s confirmed that 250 million years ago (see Figure 25.2A) the continents of today were part of a single land mass. By 150 million years ago (see Figure 25.2B) this had broken into large areas of land separated by oceans. And by 50 million years ago (see Figure 25.2C) the continents were even closer to their present positions.

Earth's crust is thicker in some places (where there are continents) than in others (where there are oceans). In mid-ocean, and in some places close to continents, there are lines of weakness. These are cracks through which hot molten rock is forced up from the mantle. As it cools and solidifies submerged mountain ranges are formed.

Earth's crust comprises a number of plates which move on the semi-liquid mantle. Where molten rock is pushed up between the plates it forces them apart. This accounts for the movement of continents.

A result of the plates moving apart in some places is that they come together in others. Where plates are presssed slowly together great upfolds form mountain ranges. Where one plate slides under another deep ocean trenches are formed; and nearby there are areas of greatest volcanic activity. In other places plates rub against one another, causing earth tremors or more violent earthquakes.

25.2 Rock formation and rock destruction

You may see layers of rock in a cliff or quarry face (see Figure 25.3). When deep holes are dug, so that useful materials deep below the surface can be extracted, the hole has to be cut through layers of different kinds of rock. Similarly, when geologists bore even deeper holes and extract long cylinders of rock, rather like removing a cylinder of apple with an apple corer, the core of rock comprises many layers of different kinds of rock. Such deep bore holes may be made, for example, so that underground water, oil or gas can be extracted. They may also be made as part of a geological survey, just to study the nature of the rocks in any area or — as test drillings — in search of new sources of materials that may be used by people.

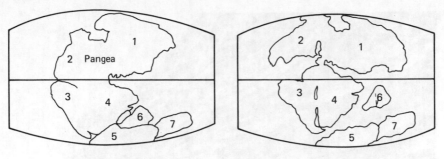

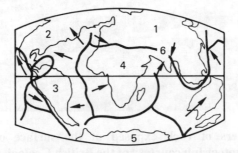

Figure 25.2 the movement of continents. Key: (1) Eurasia; (2) North America; (3) South America; (4) Africa; (5) Antarctica; (6) India; and (7) Australia and New Zealand. Thick lines indicate plate boundaries; arrows indicate directions in which plates are moving

About 90 per cent of Earth's crust is formed of *igneous rocks* (Latin *igneus* = fiery, hence the English word ignite). These rocks are formed when hot fluid, called magma, from the mantle (see Figure 25.4), is forced up through cracks in the crust. If the magma flows as lava from a volcano on land, or spreads from cracks in the sea bed, it forms a new surface layer of rock. Because this cools quickly it is composed of small crystals and is called basalt. However, if magma is forced between deeper layers of rock it is insulated by the overlying rock. Then it cools slowly, forming larger crystals, and is called granite.

**Figure 25.3 layers of rock exposed at the surface of Earth's crust.
(Photograph courtesy of the British Geological Survey.)**

Investigation 25.1 *Regaining a pure solute from a solvent by*
crystallisation

You will need a solution of sodium chloride in water, two evaporating
dishes, two tripods, two wire gauzes, a bunsen burner, a glass rod and
eye protection.

Proceed as follows
1 Arrange an evaporating dish containing some of the salt solution
   on a gauze on one of the tripods, over a bunsen burner.
2 Boil the solution to evaporate some of the water.
3 Test the solute from time to time by allowing one drop to cool at
   the end of the glass rod. When you see crystals in this drop, turn off
   the bunsen.
4 Note the size of the crystals formed. These crystals can be
   separated from the remaining solvent by filtration.
5 Arrange the other evaporating dish as in step 1. Wearing eye
   protection boil off all the water. Compare the size of the crystals
   formed quickly, with the size of those formed slowly.

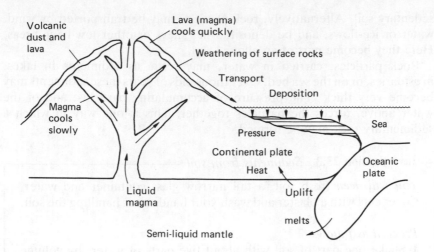

Figure 25.4 rock formation and rock destruction. Continental and oceanic plates of Earth's crust represented on the semi-liquid mantle

---

Investigation 25.2 *Suspensions, sediments and emulsions*

You will need finely powdered chalk, oil, water and two glass jars.

Proceed as follows
1 Half fill the jar with water, add some finely powdered chalk, then shake the jar to mix the water and chalk. Chalk is insoluble in pure water but the fine particles will remain in the water for some time as a *suspension* of chalk in water. Chalk is heavier than water so the particles settle as a *sediment*.
2 Half fill the other jar with water, add some oil, then shake the jar to mix the water and oil. Oil is insoluble in water but very small oil droplets will remain in the water for some time as an *emulsion* of oil in water. Oil is lighter than water so it soon forms a layer above the water surface.

Over much of Earth's crust, igneous rocks are covered by rocks formed from sediments, called *sedimentary rocks*.

All exposed rocks, however they were formed, are broken by physical changes, including: (1) alternate expansion and contraction as surface temperatures rise and fall (see page 37), and (2) expansion of water as it freezes (see page 38) in surface cracks and crevices.

Rock particles may remain on the rocks from which they have broken away, held in cracks or by plant roots. The particles are then part of a sedentary soil. Alternatively, rock particles may be transported by wind, water or ice-flows, and be deposited in valleys or other low-lying areas. Here they become part of an alluvial soil.

Rock particles, carried in water, may settle as sediments in lakes, in estuaries, or on the sea bed. Over thousands of years these sediments may become very thick. The pressure of accumulating particles, and of the water above, forces the particles together. This is one way in which a sedimentary rock may be formed.

Investigation 25.3 *Sediments from soil*

You will need fresh soil, a tall narrow glass container and water. Cover cuts with a plaster and wash your hands after handling the soil.

Proceed as follows
1 Shake one part of soil with about five parts of water, by volume (enough to fill the tall glass container).
2 Empty the mixture into the container and allow it to stand. The heaviest particles will settle first, then the lighter particles. Any material that is lighter than the water will float.
3 Observe the sediment forming. Note that the lightest particles remain in suspension for some time but that the water gradually becomes clear.

Because the heaviest particles are deposited first, the particles in a water current are carried different distances and deposited separately. Mud banks contain mainly smaller particles, and sand banks contain mainly larger particles. Under great pressure, sediments of sand may develop into sandstone — a coarse grained rock.

Rocks can be *weathered* by chemical as well as physical action. For example limestone and chalk, which are mainly calcium carbonate, are affected by rain. They are porous and so absorb water, and are affected both at the surface and below ground. Carbon dioxide dissolves in water, forming carbonic acid:

$$CO_2 \text{ (g)} + H_2O \text{ (l)} \longrightarrow H_2CO_3 \text{ (aq)}$$
$$\text{carbonic acid}$$

This dilute acid reacts with calcium carbonate (which is insoluble in pure water), forming calcium hydrogen carbonate which dissolves in water:

$$CaCO_3 \text{ (s)} + H_2CO_3 \text{ (aq)} \longrightarrow Ca(HCO_3)_2 \text{ (aq)}$$
$$\text{Calcium carbonate} \qquad\qquad \text{calcium hydrogen carbonate}$$

Some fuels, including coal, contain some sulphur. When they are burned in air sulphur dioxide is formed:

$$S\ (s)\ +\ O_2\ (g) \longrightarrow SO_2\ (g)$$

Sulphur dioxide dissolves in water:

$$SO_2\ (g)\ +\ H_2O\ (l) \longrightarrow H_2SO_3\ (aq)$$
$$\text{sulphurous acid}$$

One result of sulphur dioxide pollution of the atmosphere, therefore, is *acid rainfall*. This also dissolves calcium carbonate, and so damages limestone buildings and marble statues, as well as weathering chalk, limestone and marble rocks.

As a result of calcium carbonate in rocks being dissolved, calcium ions are present in sea water. They are absorbed by unicellular organisms in the sea, some of which have shells of calcium carbonate. When these organisms die they decay, but the microscopic shells remain. Insoluble calcium carbonate, accumulating on the sea bed over thousands of years, forms the hard, fine-grained calcareous sedimentary rock called limestone. Calcium ions are also absorbed by marine molluscs, some of which have shells composed mainly of calcium carbonate. When these larger animals die and decay their shells remain and, accumulating over thousands of years, form the softer calcareous sedimentary rock called chalk.

About 250 million years ago, Earth's climate was much warmer than now. Some areas were covered by swampy forests of large ferns and trees. When these fell to the ground they accumulated as a layer of submerged vegetation. Later, as a result of Earth movements, these forests were completely submerged. For long periods, sediments of sand and silt settled upon and compressed the layer of vegetation. In this way, over many thousands of years, the remains of the vegetation were changed into coal — which is mostly carbon. In some coal it is possible to see impressions of leaves and other parts of plants. From these fossils we know what kinds of plants were growing long ago.

Oil and natural gas occur in Earth's crust. Oil is thought to have been formed in a similar way to coal, but in the sea and from animals not plants. When marine animals died they sank and accumulated on the sea bed. There, where there was very little oxygen, they decayed slowly, if at all. Later they were covered by other materials, and the result of heat and pressure over thousands of years was the production of oil and natural gas. These were forced up, under great pressure, through layers of permeable rock and accumulated below impermeable rock with the gas above the oil (see Figure 25.5). In oil wells it is gas pressure that forces the oil up through bore holes to the surface.

Natural gas, which is mostly methane (see page 293), is known as marsh gas because it is also formed when vegetation decays in the absence of air

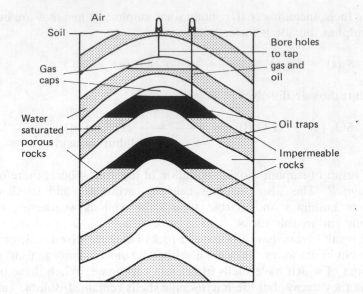

Air
Soil
Gas caps
Water saturated porous rocks
Bore holes to tap gas and oil
Oil traps
Impermeable rocks

Figure 25.5 oil and natural gas in Earth's crust, vertical section through surface layers of rock: oil and natural gas in porous rocks, trapped by layers of impermeable rocks

below the surface of swamps and marshes. Methane was produced in similar conditions when coal was formed, and the release of this trapped gas has fuelled many explosions in coal mines.

In addition to igneous and sedimentary rocks, there is a third type called metamorphic rocks. These are formed when igneous or older sedimentary rocks, deep below the surface, are changed by great heat and pressure. The term metamorphosis means change of form. Examples of metamorphic rocks are slate (formed from fine–grained clayey sediments) and marble (formed by the recrystallisation of limestone).

Other changes are shown in Figure 25.4. Note that old rocks may become part of Earth's fluid mantle; and material from the mantle may move up and solidify as new rocks in Earth's crust. Other changes result from Earth movements, raising rocks formed as sediments. All exposed rocks are weathered (physically, see page 281; and chemically, see page 282) and transported, forming new sediments that may develop into rocks. All these changes can be represented as a rock cycle (see Figure 25.6).

Since Earth was formed, about 5000 million years ago, its surface — which seems so permanent to us — has been changing all the time. This is a key principle in geology, called the *principle of uniformity: what is happening now on Earth has happened similarly in the past.*

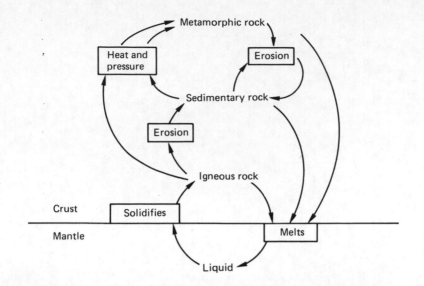

**Figure 25.6 the rock cycle: crust formation from mantle, transformation
and erosion of rock, and mantle formation from crust**

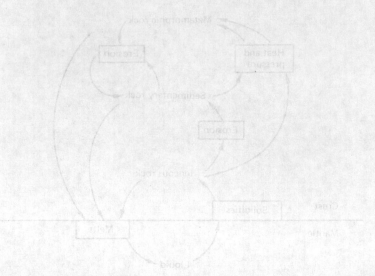

26 Material resources

A *resource* is *any source of materials or satisfaction used by people*. Air, water, rocks and living things are all sources of materials used by people. The countryside and wilderness are sources of materials and also sources of satisfaction which are sometimes called amenities or tourist resources.

26.1 Useful materials from Earth's crust

Many useful materials are extracted from rocks. Some, for example stone, gravel and sand, are extracted in a usable form. Some, for example metal ores, are called *raw materials* because, after extraction, they have to be processed to obtain useful substances. Some are useful both as they are and as raw materials. For example, coal, oil and natural gas can all be used as fuels (sometimes called fossil fuels) and they can also be used, by the chemical industry, as raw materials from which many useful substances can be obtained.

The amount of each of these useful materials in Earth's crust decreases as it is extracted. Once each source of raw materials is exhausted, the resource cannot be replaced. These are therefore called finite, or limited, or irreplaceable, or *non-renewable resources*.

To make these resources last as long as possible, people should be discouraged from just throwing things away when they have no further use for them. For example, instead of discarding scrap metals, they can be collected and reprocessed so they can be used again. This is called *recycling* (see Figure 26.1). Anything people can do to make non-renewable resources last longer is called the *conservation of resources*.

Figure 26.1 recycling materials

Extracting metals from ores

Silver and gold are exceptional metals, in that they are unreactive and occur pure in nature. Other metals react with non–metals and occur naturally as part of chemical compounds, in rocks called metal ores or minerals.

Iron is extracted from haematite, an ore containing iron oxide, in a blast furnace (so called because a blast of hot air is forced into the furnace). Hot air rises through a mixture of coke, iron ore and limestone which are added continuously at the top (see Figure 26.2).

The burning coke (mostly carbon) combines chemically with oxygen (is oxidised), producing carbon monoxide and releasing heat energy:

$$2C \text{ (s)} \quad + \quad O_2 \text{ (g)} \quad \longrightarrow \quad 2CO \text{ (g)} \quad + \quad \text{Heat}$$

| carbon | oxygen | carbon monoxide | energy |

The iron ore is decomposed by heat. It loses oxygen (is chemically reduced), releasing more heat energy. The oxygen combines with the carbon monoxide (a reducing agent) forming carbon dioxide.

$$Fe_2O_3 \text{ (s)} \quad + \quad 3CO \text{ (g)} \quad \longrightarrow \quad 2Fe \text{ (s)} + 3CO_2 \text{ (g)} \quad + \quad \text{Heat}$$

| iron (III) oxide | carbon monoxide | iron | carbon dioxide | energy |

A reaction such as this, in which one substance is reduced as another is oxidised, is called a reduction-oxidation or redox reaction.

As a result of the heat energy released in these chemical reactions, the temperature in the centre of the furnace is about 1500°C and this decomposes the limestone (calcium carbonate):

$$CaCO_3 \text{ (s)} \quad \xrightarrow{\text{Heat energy}} \quad CaO \text{ (s)} \quad + \quad CO_2 \text{ (g)}$$

| calcium carbonate | | calcium oxide | carbon dioxide |

The calcium oxide, with sandy impurities from the iron ore, forms a molten calcium silicate slag:

$$CaO \text{ (s)} \quad + \quad SiO_2 \quad + \text{ Heat} \longrightarrow \quad CaSiO_3 \text{ (l)}$$

| calcium oxide | sand | | calcium silicate |

This liquid slag is run off the surface of the liquid iron (see Figure 26.2).

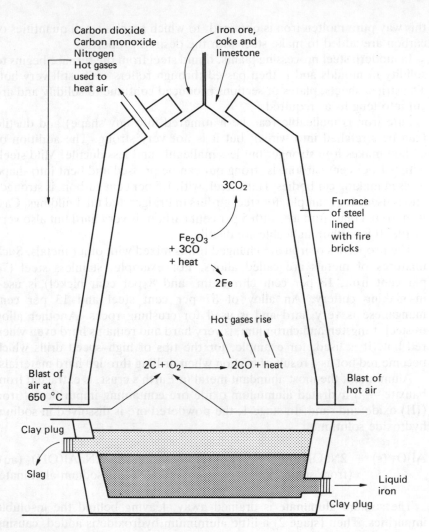

Carbon dioxide
Carbon monoxide
Nitrogen
Hot gases
used to
heat air

Iron ore,
coke and
limestone

$3CO_2$

Fe_2O_3
$+ 3CO$
$+ heat$

$2Fe$

Hot gases rise

$2C + O_2 \longrightarrow 2CO + heat$

Furnace
of steel
lined
with fire
bricks

Blast of
air at
650 °C

Blast of
hot air

Clay plug

Slag

Liquid
iron

Clay plug

Figure 26.2 the decomposition of iron oxide by heat in a blast furnace, to extract iron

Liquid iron from the blast furnace, containing about 5 per cent carbon, is tapped continuously. This can be poured into moulds (cast) and allowed to solidify in blocks (ingots) of cast iron (also called pig iron). Alternatively, to make steel, the liquid iron from the blast furnace can be transferred directly to a steel furnace. This saves the cost of having to reheat the cast iron to make steel.

Steel scrap can be added to the liquid iron in the steel furnace, for reprocessing. Oxygen is injected into the furnace to burn off the carbon remaining in the liquid iron from the blast furnace and in the scrap steel. In this way pure molten iron is obtained, to which precise small quantities of carbon are added to make steel with the desired properties.

In modern steel processing plants, liquid steel from the furnace begins to solidify in moulds and is then passed through rollers while still very hot. The strips, sheets, plates or sections produced continue to solidify and are cut into lengths as required.

Pure iron is malleable (can be hammered into any shape) and ductile (can be stretched into wires), but it is not very strong. The addition of carbon makes iron stronger but less malleable and less ductile. Mild steel, with 0.1 per cent carbon, is strong but can be pressed and bent into shape — as in making car bodies. Hard steel, with 1.5 per cent carbon, is stronger and is used, for example, for steel girders in bridges and tall buildings. Cast iron from a blast furnace, with 5 per cent carbon, is very hard but also very brittle. It is neither malleable nor ductile.

The properties of iron are changed if it is mixed with other metals. Such mixtures of metals are called alloys. For example, stainless steel (74 per cent iron, 18 per cent chromium, and 8 per cent nickel) is used in making cutlery. An alloy of 87 per cent steel and 13 per cent manganese is very hard and is used for crushing rocks. Another alloy of steel, tungsten and chromium, is very hard and remains hard even when red hot. It is used, for example, for the tips of high–speed drills which become red hot as a result of friction when drilling through hard materials.

Aluminium, the most abundant metal in Earth's crust, is extracted from bauxite — a hydrated aluminium oxide ore containing impurities of iron (III) oxide and sand. In stage 1, the powdered ore is dissolved in sodium hydroxide solution:

$$Al_2O_3 \ (s) \ + \ 2NaOH \ (aq) \ + \ 3H_2O \ (l) \longrightarrow 2NaAl(OH)_4 \ (aq)$$
$$\text{(from stage 2)} \qquad\qquad\qquad\qquad\qquad \text{sodium aluminate}$$

The sodium aluminate is drained away, leaving behind the insoluble impurities. Then (stage 2) a little aluminium hydroxide is added, causing more to be formed as a precipitate (a solid suspended in a solution):

$$NaAl(OH)_4 \ (aq) \ \xrightarrow[\text{Al(OH)}_3]{\text{some}} \ NaOH \ (aq) \ + \ Al(OH)_3 \ (s)$$
$$\qquad\qquad\qquad\qquad\qquad\qquad\qquad \text{(to stage 1)}$$

The aluminium hydroxide precipitate obtained is heated (stage 3):

$$2 \ Al(OH)_3 \ (s) \longrightarrow Al_2O_3 \ (s) \ + \ 3H_2O \ (l)$$
$$\text{aluminium hy-} \qquad\qquad \text{aluminium} \qquad \text{water}$$
$$\text{droxide} \qquad\qquad\qquad \text{oxide}$$

Aluminium oxide, with a melting point of 2000 °C, is mixed with cryolite, an ore containing sodium aluminium fluoride (stage 4). This mixture, which has a lower melting point (950 °C), is melted and used as an

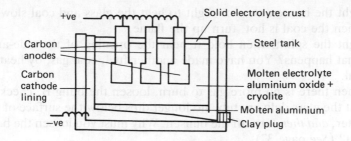

Figure 26.3 the decomposition of aluminium oxide by electrolysis to extract aluminium

electrolyte (a liquid that conducts electricity, see page 158). The mixture remains molten (at about 1000 °C) as a result of the heating effect of the electric current (see page 160). The electric current also has a chemical effect: the electrolysis (electrolytic splitting) of aluminium hydroxide (see Figure 26.3). Both electrodes (cathode and anode) are made of graphite (carbon).

At the anode: $2O^{2-} \longrightarrow 4e + O_2$ (g)

Then: O_2 (g) $+ C$ (s) $\longrightarrow CO_2$ (g) $+$ Heat

The heat energy released when the oxygen and carbon combine, forming carbon dioxide, also helps maintain the high temperature required to keep the mixture molten.

At the cathode: $\quad Al^{3+} \qquad + \qquad 3e \qquad \longrightarrow \qquad Al$ (s)

$\qquad\qquad$ aluminium ions $\quad$ electrons $\qquad\qquad\qquad$ aluminium

Extracting chemicals from coal and from crude oil

Crude oil, as it comes from oil wells, is a mixture of many chemicals. So is coal.

Investigation 26.1 *Obtaining chemicals from coal*

You will need the apparatus illustrated in Figure 26.4, coal dust, distilled water, pH indicator paper, a splint and eye protection.

Proceed as follows

Warning Many severe cuts have been caused by glass breaking as it is pushed through a rubber bung. The apparatus should therefore be prepared by a teacher or technician, not by a student. You must wear eye protection, always, when heating glass or chemicals.

1 Light the bunsen at a low light to heat the glass and coal slowly.
2 When the coal is hot, turn up the flame.
3 Light the splint, then hold it near the opening of the side-arm. What happens? You have made another fuel, coal gas, by heating coal.
4 When there is no more gas to burn, loosen the bung B, checking that the glass delivery tube no longer dips below the surface of the water, *and then* turn off the bunsen. Why must you loosen the bung first? (*See* page 37).
5 Use the indicator paper to test the pH of the clear liquid in the tube. This is now an ammoniacal liquor which could be used to make a fertiliser or household cleaning fluid.
6 Note the oily material on this liquor. This is tar, which could be used as tar or distilled as a source of other chemicals.
7 When the apparatus is cool, examine the solid remaining in the glass tube. This is coke, used in smelting iron (see page 288).

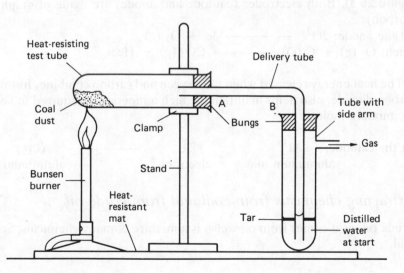

Figure 26.4 chemicals from coal

In an oil refinery, crude oil is heated to about 350°C. At this temperature it is mainly a mixture of gases. These condense, as they cool, in a fractionating column, and are collected as fractions (see Figure 26.5 and page 86). Some of the gas remaining at the top of the column is used as a fuel in the refinery (to heat the crude oil) and the rest is sold as bottled gas.

The chemicals derived from coal, and from crude oil, are *hydrocarbons* (compounds of hydrogen and carbon). Natural gas is mainly methane, a

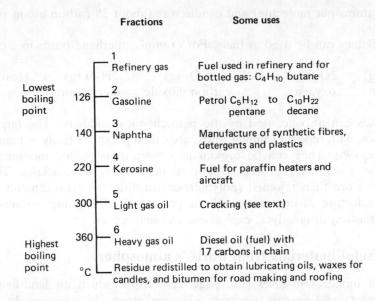

	Fractions	Some uses
	1 Refinery gas	Fuel used in refinery and for bottled gas: C_4H_{10} butane
Lowest boiling point 126	2 Gasoline	Petrol C_5H_{12} to $C_{10}H_{22}$ pentane decane
140	3 Naphtha	Manufacture of synthetic fibres, detergents and plastics
220	4 Kerosine	Fuel for paraffin heaters and aircraft
300	5 Light gas oil	Cracking (see text)
360 Highest boiling point °C	6 Heavy gas oil	Diesel oil (fuel) with 17 carbons in chain
	Residue redistilled to obtain lubricating oils, waxes for candles, and bitumen for road making and roofing	

Figure 26.5 the fractional distillation of petroleum (crude oil). (Based on Critchlow, P. Mastering Chemistry)

hydrocarbon comprising one carbon atom combined with four hydrogen atoms per molecule (formula CH_4). This can be represented diagrammatically as:

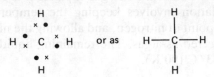

Methane and longer chain-like molecules with a similar structure, are called alkanes. For example:

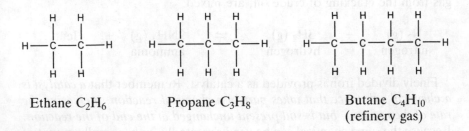

Ethane C_2H_6 Propane C_3H_8 Butane C_4H_{10}
                                                                  (refinery gas)

Other alkanes, obtained by refining crude oil, are the fractions sold as petrol (containing 5–10 carbon atoms per molecule), diesel oil (about 17 carbon atoms per molecule) and candle wax (about 28 carbon atoms per molecule).

All alkanes can be used as fuels. For example, methane burns in air:

$$CH_4 \text{ (g)} + 2O_2 \text{ (g)} \longrightarrow CO_2 \text{ (g)} + 2H_2O \text{ (g)} + \text{Heat}$$

methane   oxygen   carbon dioxide   water vapour

Alkanes can also be used by the petrochemical industry. The larger molecules, with repeated similar units, are called polymers (poly = many, see page 196). They can be broken into basic units, called monomers (mono = one). This process is known in industry as cracking. The monomers are then rejoined (polymerised) in different arrangements in the manufacture of different synthetic polymers — including manmade fibres, plastics, drugs, dyes, explosives, and detergents.

26.2 Useful materials from Earth's atmosphere

Air is a mixture of gases (see page 115), from which all land-living organisms extract oxygen (see page 168), and from which in sunlight all green plants extract carbon dioxide (see page 166).

Industrial extraction of useful gases from air

Air is cooled to $-60\,°C$, at which temperature solid carbon dioxide and ice can be removed. The remaining mixture, mainly nitrogen and oxygen, is cooled to nearly $-200\,°C$ by compressing the gas and then releasing the pressure, repeatedly. Each time the gas expands it is cooled.

Fractional distillation involves keeping the temperature at $-196\,°C$ (77 K), the boiling point of nitrogen, and allowing the nitrogen to boil until it has all been collected as a gas. The remaining liquid is mainly oxygen, which boils at $-183\,°C$ (90 K).

Industrial production of ammonia: the Haber process

Nitrogen gas, obtained by fractional distillation of liquid air, and hydrogen gas from the cracking of crude oil, are mixed:

$$N_2 \text{ (g)} + 3H_2 \text{ (g)} \rightleftharpoons 2NH_3 \text{ (g)} + \text{Heat}$$

nitrogen   hydrogen   ammonia

Finely divided iron is provided as a catalyst. Remember that *a catalyst is a chemical substance that takes part in a chemical reaction, increasing the rate of the reaction, but is still present unchanged at the end of the reaction.* Because the same chemical can be used repeatedly, only a small amount of the catalyst is needed to catalyse the production of a large amount of ammonia.

The arrows $\rightleftharpoons$ indicate that a reaction is reversible. If left alone a state of dynamic equilibrium would be established, with the different chemicals present in constant proportions while the forward and backward reactions continued.

Whatever a chemist does to change an equilibrium mixture, these changes in the mixture will tend to have the opposite effect. This summary of observations, suggested by Le Chatelier in 1885, is known as Le Chatelier's principle. For example, in the Haber process the yield of ammonia can be increased — and the rate of production — by increasing the concentration of the reactants (nitrogen and hydrogen) in the reaction chamber, and by removing some of the products (ammonia and heat energy derived from chemical energy in the reaction).

The amount of nitrogen and hydrogen in the reaction chamber is increased by compressing these gases (to 250 atmospheres), and heat from the reaction chamber containing the catalyst (see Figure 26.6) is used to heat the incoming gases to maintain a temperature of 500 °C in the reaction chamber. At the same time the outgoing mixture of ammonia, nitrogen and hydrogen gases is cooled (see heat exchanger in Figure 26.6) and then further cooled in a refrigeration unit to −40 °C. The liquid ammonia can then be run off and the unconverted nitrogen and hydrogen gases recycled.

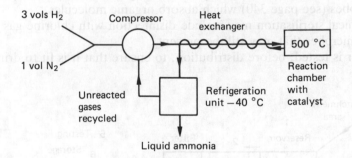

Figure 26.6 a flow diagram summarising the Haber process for producing ammonia. (Based on Critchlow, P. Mastering Chemistry)

26.3 Water as a resource

Although some organisms can exist in an almost dry state, for example the seeds of flowering plants, most organisms need water every day. We use water to drink, to promote plant growth and for many other purposes. Industrial societies, especially, collect, store and use great quantities of fresh water. Fresh water can be obtained from sea water by distillation (see page 85). Fresh water is also obtained, largely from sea water, as a result of the water cycle in nature (see Figure 24.5).

Even if it looks clean, the water in reservoirs, lakes and rivers is likely to be contaminated with chemicals that run off farm land or are present in waste-water effluents from factories and houses. It may also contain harmful microbes (see Tables 31.1 and 31.2). This is why water from lakes and rivers must be treated before it can be distributed.

Water from springs, wells and bore-holes, because it has percolated through the ground (been filtered), is likely to be fit to drink — unless it has been contaminated after percolation (for example, by things thrown into a well).

Water treatment

The stages in water treatment (see Figure 26.7) are as follows:
1 During storage in a reservoir there is usually a great reduction in the number of microbes from faeces — those that cause typhoid and other water-borne diseases die. This is due to the settlement of contaminated organic impurities, to ultraviolet rays killing the bacteria in surface waters, and to the natural death of disease-causing bacteria outside their host organisms.
2 At a purification works the water flows through grid screens into settling tanks, in which more suspended impurities settle out.
3 The water may then be passed through sand (particle size 0.5–1.5 mm) in which cleansing is due partly to filtering but mainly to saprobiotic microbes (see page 340) which absorb organic molecules.
4 Chemical sterilisation may include disinfection with chlorine gas to kill any microbes that may still be present.
5 Water is tested, before distribution, to ensure that it is fit to drink.

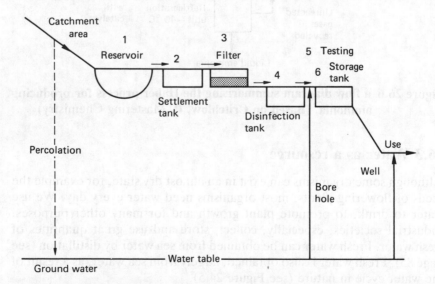

Figure 26.7 water collection, stages in treatment, and distribution: a flow diagram

6 Clean water, from storage tanks on high ground above a town, is piped directly to factories and houses. In each building it is taken directly to cold water taps (drinking water) and to a storage tank (usually in the roof space) which provides water for other uses. These tanks should be covered so the water is not fouled by the droppings of birds or rodents. However, remember that this water may become contaminated with microbes — so it should be boiled for 20 minutes if it is ever necessary to use it for drinking.

27 Magnetism and electricity

27.1 Magnetism

More than 2000 years ago mariners were using pieces of a naturally magnetic ore as lodestones (leading stones) for direction finding. These stones were mined in a part of ancient Greece called Magnesia, and so they came to be known as magnets. When a lodestone was suspended so that it was horizontal and free to move, it always rested with one end pointing north. The first steel magnets were made more than 500 years ago and used as compass needles in place of lodestones.

The scientific study of magnetism was begun by William Gilbert (see page 110) whose book *De Magnete* was published in 1600. It will help you to understand the uses of magnets if you study their properties yourself.

Investigation 27.1 *Properties of magnets*

You will need two bar magnets, thread, small objects made of iron, steel, wood and other materials.

Proceed as follows
1 Suspend one magnet so that it hangs horizontally and is free to move. Note that a magnet always comes to rest with one end pointing north. This north-seeking pole is called the magnet's north pole.
2 Hold the north pole of the other magnet next to the north pole of the suspended magnet. What happens? What happens when you hold the south poles together, and when you hold a north pole next to a south pole?
3 Hold small objects near the centre of a magnet and then near each pole. Record your observations.

Magnets are made from steel, or from alloys of iron with one or more of the metals cobalt, nickel and aluminium, or from ferrites (iron oxide with small amounts of other metallic oxides). Because objects made of iron, steel, cobalt, nickel or ferrites are attracted to magnets, these materials are called ferromagnetic materials.

Note. (1) Magnetic materials are attracted more strongly to the poles than to other parts of a magnet, (2) magnetic poles always occur in pairs (one north and one south) which are of equal strength, and (3) *like poles repel and unlike poles attract*. This last statement is *the law of magnetic poles*.

Investigation 27.2   *Magnetic fields*

You will need two bar magnets, some iron filings and a piece of thin white card.

Proceed as follows
1 Place the card flat upon one of the magnets.
2 Sprinkle a thin layer of iron filings over the card.
3 Tap one edge of the card gently and observe that the iron filings become arranged in lines.
4 Draw a diagram showing the position of the magnet (below the card) and of the lines of iron filings (on the card).
5 Pour the iron filings back into their container, then repeat this investigation with two magnets arranged in different positions.

The space around a magnet, within which the magnet exerts magnetic forces of attraction and repulsion, is called the magnet's magnetic field. In your investigations the lines of iron filings mark the positions of invisible lines of magnetic force in the magnetic field. Note that these lines of force are closest together where the magnetic force is strongest; that the lines of force never cross one another; and that each line of force extends between a north pole and a south pole. Compare your drawings with the diagrams in Figure 27.1.

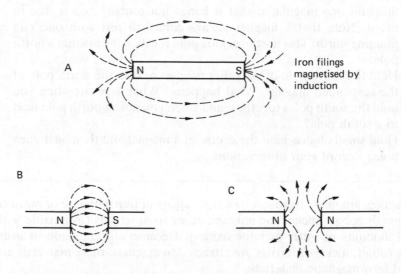

Figure 27.1 magnetic fields: (A) around a bar magnet; (B) between unlike poles; and (C) between like poles

── **Investigation 27.3** *Making a magnet* ──────────

You will need a magnet, some iron filings, a compass and a steel knitting needle.

Proceed as follows
1 Stroke the steel needle with the magnet about 20 times, always in the same direction and with each return movement in a wide circle above the magnet.
2 Dip the steel needle in the iron filings. If the steel has become a magnet (been magnetised), the filings will stick to the poles of this new magnet.
3 Point one end of the steel bar at each pole of the compass needle. The north pole of the new magnet will repel the north pole of the compass needle.

Note
1 *Repulsion is the most reliable test for magnetism.*
2 Once a steel bar has been magnetised it remains a magnet and is called a permanent magnet.
3 Another method of making a magnet is described on page 304.

The theory of magnetism

If you were to break the magnet made in Investigations 27.3 you would find, no matter how many times you broke it, that each piece would be a magnet with two poles. It is not possible to have a magnet with only one pole.

There is evidence that materials which can be magnetised are composed of groups of atoms, like small magnets, each with a north pole and a south pole. When the material is magnetised these groups of atoms become arranged with their north poles pointing in the same direction. According to this theory, when you magnetise a steel bar by stroking it with a magnet you use the magnet to pull the groups of atoms in the steel into line.

A magnet can be demagnetised by dropping it repeatedly, by heating it, or by hammering. These treatments are thought to disturb the regular arrangement, so that the groups of atoms point in all directions.

If you tried to make a magnet of iron, instead of steel, you would find that iron can be magnetised, but ceases to be a magnet as soon as the magnetising force is removed. A magnet made of iron is therefore called a *temporary magnet*. It is thought that the groups of atoms point in one direction when the magnetising force is present, but point in all directions as soon as the magnetising force is removed.

Temporary magnets are used, for example, in electric bells and buzzers, in microphones and loudspeakers, and in telephones (each of which contains a microphone and a loudspeaker). Permanent magnets are used not only as compass needles but also in electric motors and electricity generators. Another use of magnets is in tape recording — each tape is a thin plastic strip coated with finely powdered ferrites.

27.2 Electricity and magnetism

An electric current is the movement of electrons through an electrical conductor. You know these negatively charged particles move from the negative terminal of an electric cell, through a conductor, towards the positive terminal (see page 155). This movement of electrons in one direction in a conductor is called a direct current (d.c.).

When electric currents were first studied, before electrons had been discovered, the convention adopted was that electricity flowed from the positive terminal, through a conductor, towards the negative terminal. Rather than change the rules that scientists and engineers had been working to for years, this convention is still used in all practical investigations with electricity. This is why, when arrows are drawn on circuit diagrams (for example, in Figure 27.4 and Figure 27.7), they indicate the conventional current direction (not the actual movement of electrons).

In 1820 a Danish scientist, Hans Christian Oersted, noticed that an electric current in a wire caused the needle of a nearby magnet to move (as in Investigation 14.6). Now you can learn more.

Investigation 27.4   *The magnetic effect of an electric current*

You will need a circuit arranged as in Figure 27.2 (using a 4.5v battery) and a compass.

Proceed as follows
1 Arrange the wire over the compass, parallel to the compass needle.
2 Switch on the electric current and observe the movement of the compass needle. Record the direction of this movement.
3 Switch off the current and observe the compass needle. Record your observation.
4 Reverse the battery connections, so that when you switch on the current it is in the opposite direction. Record your observations as in steps 2 and 3.
5 Rearrange the wire under instead of over the compass needle. Repeat steps 2 and 3.
6 Study your records. Can you detect a pattern in your observations?

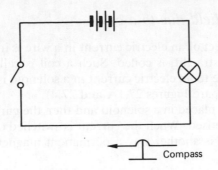

Compass

Figure 27.2 the magnetic effect of an electric current. Indicator bulb glows when current flows, and the compass needle is deflected to one side

When an electric current flows in a conductor there is a magnetic field around the conductor. This can be studied by passing the conductor through a horizontal white card, and using a small compass (a plotting compass) to indicate the position of lines of force. The direction in which the north pole of the compass needle points is called the direction of a line of force. Note, in Figure 27.3, that the lines of force are in concentric circles and they all point in the same direction (called the direction of the magnetic field). If the direction of the electric current in the conductor were now reversed, the direction of the field would also be reversed. Knowing the direction of the current you can predict the direction of the field. Imagine you are driving a right-handed screw in the direction of the current, the lines of force are then in the same direction as you are turning the screw (see Figure 27.3).

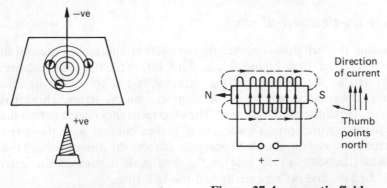

−ve

+ve

Direction of current

N   S

Thumb points north

Figure 27.3 plotting lines of force with small compasses

Figure 27.4 magnetic field around a coil; and the right-hand grip rule

The magnetic field due to a coil

The magnetic effect of an electric current in a wire is increased if the wire, instead of being straight, is coiled. Such a coil is called a solenoid. The magnetic field due to an electric current in a solenoid is similar to that of a bar magnet (compare Figures 27.1A and 27.4).

If a steel rod is placed in a solenoid and then the current is switched on, the steel is magnetised. When the current is switched off the steel remains magnetised. This is another way a permanent magnet can be made (see also page 301).

When a solenoid is wound on a core of soft iron the magnetic effect is stronger than if a steel rod is used as the core, but the iron ceases to be a magnet as soon as the electric current is switched off. A solenoid with a core of soft iron is therefore called an *electromagnet: it is a magnet only when the current is switched on*. It is a temporary magnet.

The strength of an electromagnetic can be increased: (1) by increasing the current flowing through the wire, (2) by increasing the number of turns of wire in the coil, and (3) by using a magnet in which the poles are close together — as in a C- or U-shaped magnet.

The direction of the magnetic field around a solenoid or electromagnet depends on the direction of the current in the solenoid. You can use the right–hand screw rule (see page 303) to predict the direction of the lines of force around each wire. Alternatively, if you imagine that you are gripping the solenoid with the fingers of your right hand pointing in the direction of the current, then your thumb will point to the north end of the magnetic. This is the right–hand grip rule.

Note that the coils at the ends of a U-shaped electromagnet (see Figure 27.5) are wound in opposite directions: clockwise near the south pole and anticlockwise near the north pole.

How an electric bell works

Pressing the bell-push switches on the current and makes the coil into an electromagnet (see Figure 27.5). The soft iron bar (the armature) is attracted to the magnet and the striker hits the gong. As the armature moves towards the magnet it loses contact with the screw. This breaks the circuit, switching off the current. The electromagnet ceases to be a magnet, so the armature springs back. As it makes contact with the screw, the current flows again, the electromagnet attracts the armature, and the gong strikes the bell. For as long as the bell-push is pressed the current is switched on and off repeatedly and the bell rings.

The force on a conductor of electricity in a magnetic field

A wire carrying an electric current is surrounded by a magnetic field (see Figure 27.3). You might expect, therefore, that such a wire would be

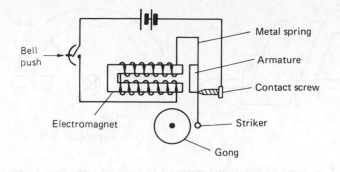

Figure 27.5 an electric bell

affected by the field of a nearby magnet — and so it is. For example, if a wire that is part of a circuit passes through a magnetic field (as in Figure 27.6) and the current is switched on, a force will act on the wire. As a result, the wire will move if it is free to do so. With the arrangement illustrated in Figure 27.6, the wire would move upwards.

The directions of (1) the current, (2) the magnetic field, and (3) the movement of the wire, are at right angles to one another (as are the walls and floor at the corner of a room). You can remember these directions from the left–hand rule. Arrange your thuMb and First and seCond fingers of your left hand so that they are at right angles to one another (as in Figure 27.6). If your First finger points in the direction of the magnetic Field, and your seCond finger in the direction of the Current, then your thuMb will point in the direction of the Movement of the wire.

Figure 27.7A represents a loop of wire placed in a magnetic field so that the wire is free to move. When the current is switched on, the forces acting on the sides of the wire loop are in opposite directions. Use the left-hand rule to check this. These forces will twist the loop until it is at right angles to the magnetic field. This effect is used in many electrical measuring instruments (for example, the moving coil galvanometer) and in electric motors.

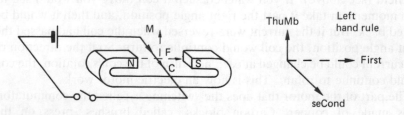

Figure 27.6 wire free to move in a magnetic field (F), when an electric current (C) flows in the wire

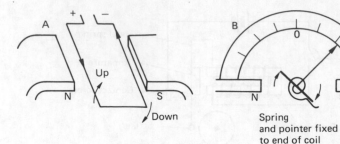

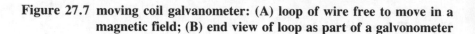

Figure 27.7 moving coil galvanometer: (A) loop of wire free to move in a magnetic field; (B) end view of loop as part of a galvonometer

A moving coil galvanometer

A galvanometer (named after Galvani, see page 153) is an instrument used for measuring small d.c. electric currents.

It consists of a flat coil pivoted in the magnetic field of a permanent magnet (see Figure 27.7B). A pointer is fixed to the end of the coil, so that a movement of the coil moves the pointer across a scale. This happens when an electric current flows through the coil.

A spring is attached to the coil to oppose the movement. The greater the current passing through the coil, the greater the force acting against the spring and the further the pointer moves across the scale. When you connect a moving coil meter in a circuit, you must ensure that the current flows through the meter in the correct direction. If the current were reversed the coil would move in the opposite direction, the spring would not then restrain the movement and the meter could be damaged permanently.

An electric motor

When an electric current is passed through a coil that is pivoted in a magnetic field, the coil turns until the plane of the coil is at right angles to the field (see above). If you watched such a coil move you would see its own momentum take it past the right angle position, and then it would be pulled back. But if the current were reversed when the coil had passed the right angle position, the coil would continue to turn. So if the direction of the current could be changed at each half turn of the coil's rotation, the coil would continue to rotate. This is how an electric motor works.

The part of the motor that does the switching is called the commutator. It is made of copper. Carbon blocks, called brushes, press on the commutator to maintain the electricity supply to the coil. The commutator turns with the coil and ensures that the side of the coil that is moving up, past the south pole of the magnet (see Figure 27.8) is always connected to the same electrode of the battery.

This type of electric motor is called a d.c. motor because it works only when connected to a direct current supply. However, a motor as simple as the one described (see also Figure 27.8) would not work smoothly. It would have its maximum turning effect only twice in each revolution, as the coil cut through the lines of force between the two poles of the magnet. This is why, instead of one moving coil, a d.c. motor is made with many separate coils wound in such a way that as they turn there is always one coil exerting a maximum turning effect.

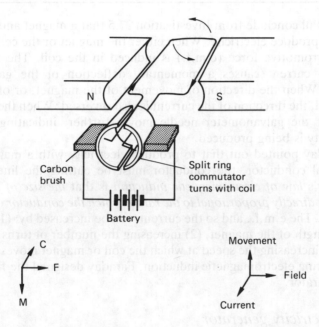

Figure 27.8 a d.c. motor. N and S = poles of magnet; C = current; F = magnetic field; and M = movement of wire. Arrows indicate directions of current flow, magnetic field, and rotation of wire loop

27.3 Generating electricity

After Oersted's discovery, in 1820, that an electric current flowing through a conductor produces a magnetic effect (see page 302), scientists reasoned that it should be possible to reverse this process and use a magnet to produce electricity in a conductor. In 1831 Faraday showed how this could be done.

Investigation 27.5 *Producing electricity using a magnet and coil*

You will need a strong permanent magnet, a coil with many turns of insulated copper wire, and a galvanometer (with zero at centre).

Proceed as follows
1 Connect the coil to the galvanometer.
2 Watch the galvanometer as you: (a) insert the magnet in the coil; (b) hold it steady; (c) pull it out; and (d) put it in and out at different speeds. Then repeat with the magnet reversed.
3 Keep the magnet still and move the coil over the magnet.

You will conclude from Investigation 27.5 that a magnet and coil can be used to produce electricity. When either the magnet or the coil is moving, an electromotive force (e.m.f.) is induced in the coil. The flow of the induced current causes a momentary deflection of the galvanometer needle. When the direction of movement of the magnet, or of the coil, is reversed, the direction of the current is also reversed. When the movement is faster the galvanometer needle moves further, indicating that more electricity is being produced.

Faraday pointed out that to produce electricity with a magnet and an electrical conductor, the conductor must be cutting the lines of force. Faraday's *law of electromagnetic induction* is that *the size of the induced e.m.f. is directly proportional to the rate at which the conductor cuts the lines of force*. The e.m.f., and so the current, can be increased by: (1) increasing the strength of the magnet, (2) increasing the number of turns on the coil, and (3) increasing the speed at which the coil or magnet moves. Soon after discovering electromagnetic induction, Faraday designed the first electricity generator.

An electricity generator

In a generator the coil cuts through magnetic lines of force repeatedly. In Figure 27.9 the coil of a simple generator is represented by one loop of copper wire, attached at its ends to two copper slip rings. Carbon brushes, pressing on these slip rings, are connected to a galvanometer. There is a complete circuit of conductors.

An e.m.f. is induced in the coil as it rotates, cutting the lines of magnetic force repeatedly. The direction of the induced current can be predicted. Consider first — using the left-hand rule (see page 305) — the direction in which a current would have to flow to produce a clockwise rotation of the coil in a motor. In a generator, a clockwise movement of the coil will produce a current in the opposite direction.

When the coil is moving through the horizontal position (as in Figure 27.9A) it is cutting the most lines of force. The induced current is therefore

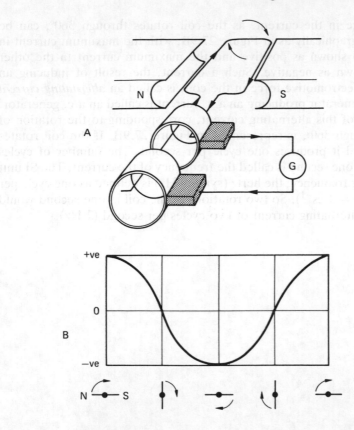

Figure 27.9 (A) a simple generator; and (B) diagram representing an a.c. current and end views of coil.

at a maximum. Note that one side of the coil is moving up and the other side is moving down, so the induced e.m.fs act in opposite directions — causing a current through the circuit as a whole in one direction (as indicated by the arrows on the wire in Figure 27.9A).

When the coil is moving through the vertical position its wires are not cutting the lines of magnetic force, so there is no induced e.m.f. and no current flows. That is to say, the induced current has changed from maximum to zero in a quarter of a rotation of the coil.

As the coil moves from the vertical position an e.m.f. is again induced, reaching a maximum when the coil is horizontal. But note that the induced e.m.fs are now in the opposite direction, so the direction of the current through the circuit is reversed.

As the rotation of the coil continues, the current again falls to zero when the coil is vertical, changes direction, and then rises again to a maximum as the coil moves through the horizontal position — completing one rotation.

This change in the current, as the coil rotates through 360°, can be represented graphically as in Figure 27.9B, with the maximum current in one direction shown as positive and the maximum current in the other direction shown as negative. Such a current, the result of inducing an alternating electromotive force in the coil, is called an *alternating current* (a.c.). The generator producing an a.c. current is called an a.c. generator.

One cycle of this alternating current, corresponding to the rotation of the coil through 360°, is represented in Figure 27.9B. If the coil rotates once a second it produces one cycle per second. The number of cycles completed in one second is called the frequency of the current. The SI unit for measuring frequency, the hertz (symbol Hz) is defined as one cycle per second ($1 \text{ Hz} = 1 \text{ cs}^{-1}$). So two rotations of the coil in one second would produce an alternating current of two cycles per second (2 Hz).

(28) Using electricity

28.1 Measuring electricity

To measure the amount of water flowing through a pipe you do not attempt to count the water molecules. The quantity of water flowing is measured in cubic metres per second. Similarly, when measuring an electric current in a wire scientists do not attempt to count the millions of electrons passing along. They measure the rate of flow of electric charge (electrons) in quantities, called coulombs, passing one point in a circuit each second.

The international (SI) unit used to measure electric current is the ampere, usually called the amp (symbol A). *A current of one ampere is a rate of flow of one coulomb of electric charge per second* (about 6.28×10^{18} electrons per second).

To measure the current flowing through a component in an electric circuit (for example, through the resistor in Figure 28.1), the circuit is broken to include an ammeter. That is to say, the component and the ammeter are in series (see page 159). An ammeter reading of 0.2 A indicates a flow of 0.2 coulombs of electric charge passing though the ammeter and through the component in each second. The movement of the pointer across the scale is due to the magnetic effect of the electric current (see page 306).

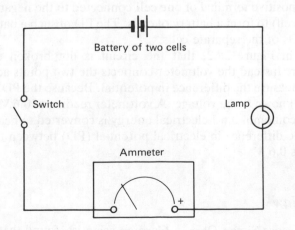

Figure 28.1 use of ammeter to measure current flowing through a lamp (a resistor)

Electrons are given energy — are caused to move through the conductor — by an electric cell (see page 162), or by some other source of electrical energy connected in the circuit. This electrical energy, the energy of moving electrons, can be converted to other forms of energy as the electrons flow around the circuit. For example, in an electric light bulb electrical energy is converted to heat energy and light energy. Note, however, that the number of electrons arriving back at an electric cell is identical with the number leaving the cell, and the number of electrons in the conducting wires and components that make up the circuit does not change.

Energy is always measured in joules (see page 108) and the amount of electrical energy being converted into other forms of energy is measured in joules per coulomb. In the international (SI) system the unit of measurement is called the volt (symbol V).

One volt is one joule per coulomb.

The instrument used to measure the change in energy of the coulombs, to other forms of energy, is called a voltmeter. In the circuit diagram (Figure 28.2) a voltmeter is being used to measure the change in energy of the coulombs (the energy dissipated) between their entering and leaving a resistor.

An electric cell is a source of energy. That is to say, it provides the force that drives electrons round a circuit. This force, called voltage or electromotive force (e.m.f., symbol E) results in a potential difference between the ends of a circuit. This difference in potential is a difference in energy level, due to the excess of electrons at the negative electrode relative to the positive electrode of the electric cell.

The potential difference (PD) of one dry cell is 1.5 V: it produces 1.5 J energy per coulomb of charge. To increase the e.m.f., and so provide a larger PD, using dry cells, two or more cells are connected in series (that is, with the positive terminal of one cell connected to the negative terminal of the next cell) to form a battery of cells. The PD of such a battery is the sum of the PDs of the separate cells.

Note, in Figure 28.2, that the circuit is not broken to include the voltmeter. Instead the voltmeter connects the two points across which we wish to measure the difference in potential. Because the PD is measured in volts it is known as the voltage. A voltmeter reading of 0.6 V indicates that, for each coulomb, 0.6 J electrical energy is converted to heat energy. As a result, the difference in electrical potential (PD) between the ends of the resistor is 0.6 V.

Ohm's law

In 1827 Georg Simon Ohm, a German scientist, found that providing the temperature of a metallic conductor did not change, *the current flowing through the conductor was directly proportional to the potential difference* between its ends. This statement, known as Ohm's law, can be written:

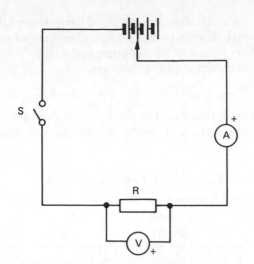

Figure 28.2 arrangement of equipment for Investigation 28.1. Note that terminals of ammeter and voltmeter marked + (or coloured red) must be connected to the +ve terminal of the battery (see page 306)

PD (V) in volts = current (I) in amperes multiplied by a constant

Every conductor offers some resistance to the flow of electricity — and this constant, the conductor's resistance (R) in ohms (symbol Ω, the Greek letter omega), is calculated: $V = I \times R$. You can see from this that, for a given voltage, the greater the resistance the smaller the current flowing.

The resistance of a conductor, such as a piece of wire, depends on its length, its thickness (diameter), the material of which it is composed, and its temperature. Resistors are components specially made to have a fixed resistance, at a particular temperature, so that they can be used to reduce the current in a circuit to the value required.

Investigation 28.1   *Demonstrating Ohm's law*

You will need the equipment represented in the circuit diagram (Figure 28.2) including 4×1.5 v cells.

Proceed as follows
1 Prepare a table, like Table 28.1, in your laboratory notebook.
2 With four cells in the circuit, close the switch, record the meter readings in your table, then open the switch. Work quickly so the resistor does not have much time in which to get hot.
3 Continue your investigation with three, then two, and then only one cell in the circuit, recording your readings and working quickly as in step 2.

4 From your data calculate the value of the resistor (R) with one, two, three and four cells in the circuit and enter your results in your table. Within the limits of experimental error, if Ohm's law is correct the value should be a constant.

Note. With the circuit illustrated in Figure 28.2, from the readings on the ammeter and voltmeter, you could determine the resistance of any conductor.

Table 28.1   *Data sheet for investigation 28.1*

Number of cells	Current (amps)	PD (volts)	$\frac{V}{I}$
4			
3			
2			
1			

The triangle: (triangle with V, I, R) will help you to remember that:

$$V = I \times R; \quad \text{or} \quad \frac{V}{I} = R; \quad \text{or} \quad I = \frac{V}{R}$$

where V = potential difference in volts, I = current in amperes, and R = resistance in ohms.

Resistors in series and in parallel

Any conductors that offer resistance to an electric current can be connected either in series or in parallel (see page 159).

Resistors in series. When resistors in a circuit are in series (see Figure 28.3) the same current flows through each resistor in turn. The combined resistance (R) of the resistors, is therefore the sum of their individual resistances:

$$R = R_1 + R_2 + R_3$$

The current (I) flowing through these resistors can be calculated, from Ohm's law:

$$I = \frac{V}{R} = \frac{\text{Potential difference across all the resistors}}{\text{Combined resistance of the resistors}}$$

The potential difference across any one resistor can also be calculated.

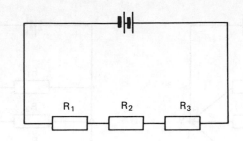

Figure 28.3 resistors (R_1, R_2 and R_3) connected in series

For example:

the PD across R_1 = V_1 = I × R_1

Example. Resistances of 4 ohms, 8 ohms and 12 ohms are connected in series to a 12 volt battery. Calculate: (a) the combined resistance (R) in ohms (Ω) of the three resistors, (b) the current (I) in amperes (A) flowing in the circuit, and (c) the PD across each bulb.

(a) Combined resistance:

$$R = R_1 + R_2 + R_3$$
$$= 4 + 8 + 12 = 24 \ \Omega$$

(b) The current flowing (I), from Ohm's Law:

$$I = \frac{V}{R} = \frac{12}{24} = 0.5 \ A$$

(c) The PD across each resistor:

$$V_1 = I \times R_1 = 0.5 \times \ \ 4 = 2 \ V$$
$$V_2 = I \times R_2 = 0.5 \times \ \ 8 = 4 \ V$$
$$V_3 = I \times R_3 = 0.5 \times 12 = 6 \ V$$

Resistors in parallel. When resistors are connected in parallel they provide extra pathways for the current (see Figure 28.4), so more current flows round the circuit than would be possible if there were only one path containing one or more resistors. In other words, putting resistors in parallel lowers the resistance of the circuit.

In Figure 28.4 each of the resistors is connected to points X and Y, so there is the same potential difference (V = voltage of battery) across each resistor. The total current (I) in this circuit is the sum of the currents flowing through the separate resistors (I_1 through R_1, I_2 through R_2 and I_3 through R_3). That is:

$$I = I_1 + I_2 + I_3$$

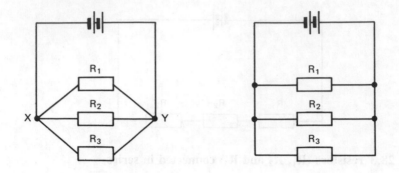

Figure 28.4 two ways of representing, in a circuit diagram, three resistors connected in parallel

Ohm's law states that $I = V \div R$, so:

$$\frac{V}{R} = \frac{V}{R_1} + \frac{V}{R_2} + \frac{V}{R_3}$$

Dividing each side of this equation by V we get the equation used to find the total resistance (R) of resistors in parallel:

$$\frac{1}{R} = \frac{1}{R_1} + \frac{1}{R_2} + \frac{1}{R_3}$$

Example. Three resistors (4 ohms, 8 ohms and 12 ohms) are connected in parallel to a 12 volt battery. Calculate: (a) the combined resistance (R) in ohms (Ω) of these three resistors, and (b) the current (I) in amperes (A) flowing through the circuit.

(a) Combined resistance (R)

$$\frac{1}{R} = \frac{1}{4} + \frac{1}{8} + \frac{1}{12}$$

$$= \frac{6}{24} + \frac{3}{24} + \frac{2}{24} = \frac{11}{24}$$

Therefore, $R = 2.18 \, \Omega$

Note. The total resistance of resistors in parallel must be less than the value of the lowest resistor when used by itself, because additional resistors in parallel provide additional pathways for the current.

(b) Current

$$I = \frac{V}{R} = \frac{12}{2.18} = 5.5 \, A$$

You could obtain the same answer by applying Ohm's Law to calculate the current flowing through each resistor and then adding the results of your calculations to find the total current.

28.2 Using electricity safely

Electricity supply to the home

In most countries the electricity supply to homes is an alternating voltage (see page 310). Each home in Britain, for example, is connected to an external supply by a mains cable containing two wires. One, covered in brown (or red) insulating material, is called the live wire. This is alternately more than 240 V +ve and less than 240 V −ve, providing an alternating current (a.c.) supply. The other, covered in blue (or black) insulating material, is called the neutral wire. This is close to zero volts. So the potential difference between the live and neutral wires is about 240 volts.

Connected to the mains wires in each home, one or more ring main circuits supply the sockets for three-pin plugs; one or more circuits supply the lights; and separate circuits are needed for an immersion heater and electric cooker (see Figure 28.5). Note that these circuits are all connected in parallel. Also, each appliance is connected (in parallel) to a live wire and a neutral wire of one of these circuits. All appliances sold for domestic use in Britain must therefore be designed to work on a 240 V a.c. supply.

Fire risks due to overloading an electric circuit

All wire conductors used in electrical appliances; in leads from plugs to appliances; and in the wiring below floors, in walls and above ceilings, provide some resistance to the flow of an electric current. As a result, some electrical energy is used in overcoming this resistance and is changed to heat energy in the wire.

If a circuit is overloaded, by using (a) an appliance that requires more current than the circuit was designed to carry, or (b) too many electrical appliances at one time and taking more current than the circuit was designed to carry, the wires can get so hot that they start a fire in nearby inflammable materials. This is why each electrical circuit is protected by a fuse. These fuses are in a consumer unit next to the electricity meter (see Figure 28.5). Note that each fuse, like each switch, is in the live wire. The fuse ratings (see Figure 28.5) indicate the maximum current each circuit is designed to take. Each fuse is a thin wire that melts if the circuit is overloaded, and so breaks the circuit — switching off the electricity supply to all appliances in the circuit.

The power rating of each appliance, in watts (W), and the voltage (V) of the power supply required, are marked on the appliance. You can calculate the electric current flowing through the appliance, from the relationship:

watts = amperes × volts.

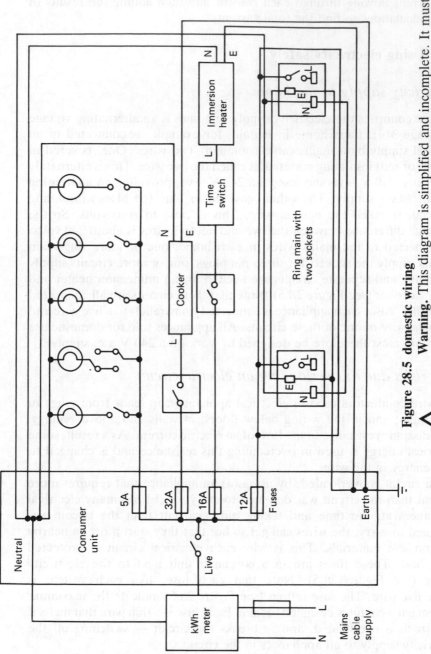

Figure 28.5 domestic wiring. This diagram is simplified and incomplete. It must not be used as a guide to electrial installation, maintenance, or repair

For example, for a light bulb labelled 240 V and 60 W, connected to a 240 V a.c. supply:

Current = 60 W÷240 V = 0.25 A.

The value of a fuse is marked on the fuse in amperes. If a fuse blows it must be replaced by a fuse of identical value (for example, replace a blown 5 A fuse with a new 5 A fuse). However, if a fuse blows because the circuit is overloaded, enough appliances should be disconnected from the supply, to reduce the load on the circuit, before the old fuse is replaced by a new one. Otherwise, the new fuse will blow when the same appliances are switched on again.

A circuit can be overloaded by using one appliance that requires more current than the circuit was designed to provide. For example, an electric heater labelled 2000 W and 240 V a.c., if connected to a 240 V a.c. supply will take a current of more than eight amps:

$$\text{Watts} = \text{amps} \times \text{volts, so } \frac{2000 \text{ W}}{240 \text{ V}} = 8.33 \text{ A}$$

If this heater were plugged into a lighting circuit it could cause a fire.

The overheating of wires in a circuit can also result if, due to worn insulation or damage, there is contact between the live wire and either the neutral wire or the earth wire. This contact makes what is called a short circuit: the wires offer very little resistance so a very large current flows (I = V ÷ R). The very large current would cause overheating, and perhaps a fire, if it were not for the presence of a fuse in each circuit (see Figure 28.5).

Electric shocks

Your body (if dry) presents quite a high resistance to the flow of electricity, so low voltage appliances such as torches, portable radios and battery operated toys are safe even for children to use. But electrical appliances and electrical wiring, connected to the 240 V mains supply, are safe only if installed and serviced by properly trained electricians — and then only if used with care.

Warning
1 Never interfere with electric wiring or put anything other than a correctly wired plug into a socket.
2 Always switch off any appliance, using the wall switch, before removing a plug or electric light bulb.
3 Never touch any electric wiring or the inside of any appliance.
4 Never use any electrical appliance if you know it is not working properly, or if the lead has worn insulation or loose wires, or if there

is any water on or near the appliance, lead, plug or socket. There should be no sockets in bathrooms, and electrical appliances should not be used in bathrooms.

You would get an electric shock (see Table 28.2), which would probably kill you, if you: (a) touched a live wire at 240 V with one hand and either a neutral wire or an earth wire with the other hand — causing a large electric current to flow through your body, across your chest; or (b) touched a live wire with one hand, and any conductor connected to earth with the other hand — causing a large current to flow through your body, across your chest; or (c) touched a live wire at 240 V with just one hand when standing in water or barefoot on a wet stone floor.

Table 28.2   *Effects of an electric shock on people*

Current (amps)	How the shock would affect you
less than 0.001	You would feel the shock.
0.001 to 0.015	Your muscles would contract, and might tighten your hold on the wire.
0.015 to 0.030	Would make breathing difficult.
0.030 to 0.050	Would soon make you unconscious.
more than 0.050	Would cause burns, and soon kill you.

If an appliance is faulty, for example if a wire is loose or broken and is touching the metal casing, the appliance could be live. Anyone touching the metal casing of a live appliance (see Figure 28.6A) would get an electric shock. To prevent this, any appliance with exposed metal parts must have an earth wire. This is connected to the metal casing of the appliance at one end, and through the plug to earth (to the ground) at the other end. A properly earthed appliance would not give you a shock, even if it were switched on when a live wire was touching the casing. Any current passing through the casing would be conducted directly to earth — through the earth wire which offers very little resistance to an electric current — not through your body which offers much more resistance (see Figure 28.6B). In this way, properly earthed appliances protect users. The surge of current also melts the fuse wire, switching off the appliance. The faulty appliance must then be repaired before the fuse is replaced.

Many electrical appliances have plastic handles. These poor conductors of electricity reduce the chances of a user getting an electric shock from a

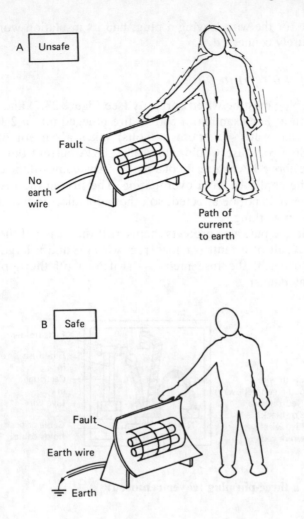

A | Unsafe

Fault

No
earth
wire

Path of
current
to earth

B | Safe

Fault

Earth wire

≡ Earth

Figure 28.6 an earth wire protects the user from electric shock

faulty appliance that is not properly earthed. Similarly, the screwdrivers
and pliers used by electricians have rubber or plastic handles to reduce the
chances of an electric shock if a live wire is touched by accident.

Some appliances, called double–insulated appliances, have a plastic
casing and no exposed metal parts. You could not get a shock by touching
the casing. These appliances must not be earthed. Examples are electric
power tools, such as electric drills. Remember, however, that you could
get a shock from any exposed live wire — for example, if the wire entering

an appliance, or the wire leaving a plug, had its insulation worn away or was not securely connected.

The use of a three-pin plug

A three-pin plug must be wired correctly (see Figure 28.7) and must have the correct fuse. For example, a 500 W fire plugged into a 240 V supply (remember that watts = amperes × volts) takes a current of $500 \div 240$ = 2.08 A. So a 3 A fuse would allow the 2.08 A current but melt if the current were above 3 A as a result of a fault — switching off the current and protecting the appliance from overheating. The fuse in the fuse box (see Figure 28.5) would not be affected, so other appliances in the same circuit would keep on working.

The fuse in the plug also protects the user. If the casing of the appliance is live as a result of a fault (or incorrect wiring), and a large current is conducted to earth, the fuse melts — switching off the appliance and removing any danger.

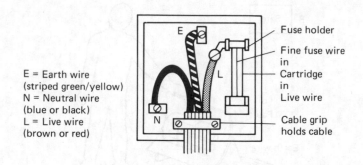

E = Earth wire
(striped green/yellow)
N = Neutral wire
(blue or black)
L = Live wire
(brown or red)

E
L
N

Fuse holder

Fine fuse wire
in
Cartridge
in
Live wire

Cable grip
holds cable

Figure 28.7 a three-pin plug (cover removed)

Paying for electricity

Energy is measured in joules (see page 108), and power (the rate of doing work, or the rate at which energy is changed from one form to another) is measured in watts (see page 113).

The unit of electrical energy consumption, recorded by the electricity meter in your home, is the kilowatt hour. One kilowatt hour is called one unit of electricity consumption and you have to pay for each unit used. When you receive a bill it states the number of units used, the cost of each unit, and the total amount you must pay.

A 1000 W (= 1 kW) electric fire switched on for one hour uses one unit of electricity. This is $1000 \times 60 \times 60 = 3\,600\,000$ joules of electrical energy (3.6×10^6 J). A 2000 W electric fire switched on for half an hour also uses one unit of electricity; and so does a 500 W fire switched on for two hours.

Investigation 28.2 *Your use of electricity*

What is the voltage of your electricity supply? Is it an a.c. or a d.c. supply? Read the labels on the appliances you use. Check that each appliance is suitable for your electricity supply. Work out the cost of using each appliance for one hour.

Energy sources

29.1 Heat energy from chemical energy

In photosynthesis, light energy is captured by chlorophyll and converted to chemical energy stored in sugars (see page 163) and in other organic molecules. When wood, peat, coal, oil and natural gas are burned this chemical energy stored in plants is converted to heat energy.

Wood

Wood was the first material used as fuel. Its use for this and other purposes has contributed to the deforestation of many countries. More trees can be grown for timber: so wood can be considered to be a *renewable resource*. However, land is needed for many other uses. The area of land available for forestry is limited: so wood is also a *limited resource*. The demand for timber for fuel and for other uses is such that deforestation in the world as a whole, is proceeding faster than re-afforestation: so woodland is a *diminishing resource*.

Existing forests should be conserved, and in many places re-afforestation is desirable. Trees contribute to the stability of soils, especially on hillsides, helping prevent soil erosion (see page 378). Woodland vegetation, with its associated soil, acts like a sponge — retaining water in upland areas and so reducing flooding on distant lowlands. For these and other reasons forests and woodlands are essential natural resources in themselves as well as being a source of useful materials.

Fossil fuels

Coal, peat, petroleum (rock oil) and natural gas were all formed from dead organisms (see page 284). They are all used as fuels (energy sources) and are therefore called fossil fuels. Do not let this word fuel mislead you. Remember that all these materials have many other uses (for example, see page 291).

The amounts of coal, peat, petroleum and natural gas on Earth are limited (see Figure 29.1A). Once used they cannot be replaced. They are *limited* and *non-renewable* resources. It might seem sensible to conserve them, and to give some thought to the needs of future generations, but the rate at which they are being used is increasing, not decreasing (see Figure 29.1B). By the year 2050 there may be no more petroleum or natural gas, and by 2350 there may be no more coal.

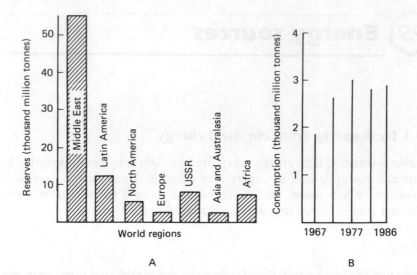

Figure 29.1 (A) world oil reserves (1986 estimates); and (B) world oil consumption in each of the years 1967, 1972, 1977, 1982 and 1986. (Based on data from the BP Statistical Review of World Energy)

The cost of fossil fuels is likely to increase, not only as the amounts remaining decrease but also because the most accessible resources were extracted first. For example, the first oil well was only 21 metres deep and in some places coal was exposed at the surface; but now some oil wells are thousands of metres deep and in many places coal, petroleum and natural gas are being extracted from below the sea bed.

At present natural gas is used mainly as an energy source (for example, for domestic heating and cooking). The materials obtained when crude petroleum is refined have many uses (see page 293).

At the start of the industrial revolution, in the eighteenth century, coal was burned to heat water, to produce steam that was needed to drive steam engines — first in factories and then to drive railway locomotives as well. Coal was also used to produce coke and town gas (see page 292). Now, many electricity power stations burn coal or oil to heat water, to produce steam needed to drive the turbines that rotate a.c. generators (alternators, see page 308).

29.2 Electrical energy from mechanical energy

There are many advantages in using other forms of energy to generate electricity. For example: (1) electricity can be carried in conductors and so does not have to be used where it is produced, and (2) when it is used, electrical energy is easily converted to other forms of energy.

Figure 29.2 a wind turbine generator (3000 kW rating at 17 metres per second) on the island of Orkney, Scotland: the 60 metre diameter rotor constructed by British Aerospace for the Wind Energy Group LS – 1 generator. (Photograph courtesy of the Wind Energy Group Ltd., Hayes, England)

Wind and water mills, with machines turned by air and water currents, were used for hundreds of years. They drove the machines used, for example, in milling flour and in making yarn and cloth. The speed at which the wheel turned varied, but the speed at which the machine turned was increased or decreased by using gears (see page 132). Wind and water mills became commercially uncompetitive in the nineteenth century following the invention of internal combustion engines, driven by energy from fossil fuels. However, as fossil fuel reserves decrease and their prices increase, scientists and engineers are being encouraged to find ways of exploiting *alternative sources of energy* that cannot be depleted — including winds and water currents.

Energy from winds

The rotating wheel of a wind mill, with blades to catch the wind, is called a rotor. New types of rotor which harness wind power more efficiently than in the past are being used to turn alternators for the generation of electricity. However, even in suitable locations (see Figure 29.2), the wind is variable and unreliable. Wind power will supplement, not replace, other energy sources.

Hydro-electricity

In the water cycle, water that evaporates at sea and ground level condenses in the clouds, and so gains potential energy. When water falls as rain and flows in rivers, this potential energy is changed to kinetic energy — which can be used to do useful work (as in a water mill).

Electricity produced from water power is called hydro-electricity. In hydro-electric power schemes, rivers in mountainous regions are dammed to form large lakes. Water stored in the lake is a store of potential energy. It can be passed along pipes, at a constant rate, to turn turbines that rotate alternators in a power station. Advantages of hydro-electricity are:

1 It can be fed into a grid system at times of peak demand, to supplement electricity generated in other ways.
2 At off-peak times the use of water and the production of electricity are easily reduced or stopped.
3 It is cheap.
4 No air pollution is caused during its production.
5 Limited resources are not depleted.

Unfortunately, few countries have a year-round rainfall and enough suitable reservoir sites for much electricity to be generated in this way.

Fossil fuels have the advantage that they can be stored and used when needed, so fossil-fuelled power stations can match peak electricity supply to demand. But at off-peak times (especially at night) supply exceeds demand. So, some hydro-electric power stations have been constructed to

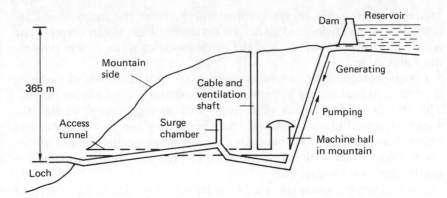

Figure 29.3 Ben Cruachan pumped storage power station in a mountain. Four 100 MW reversible turbines are used: (1) as generators, helping to meet peak demands by producing 450 million units of electricity per year; and (2) as pumps, returning water to the reservoir at off-peak times to supplement water collected from a large catchment area. (Data from North of Scotland Hydro-Electric Board)

make use of this surplus energy (see Figure 29.3). They have a dual purpose — energy production and energy storage.

Hydro-electricity can also be produced by harnessing the tremendous power of tides, ocean currents, and ocean waves. A tidal power station opened at La Rance in France in 1966. It receives water from a dam that traps the 13 metre tide. For energy stored in this way to be used, a dam has to be constructed across a bay or estuary and a tidal range of at least 5 metres is needed. The water can then be allowed to run through pipes and used to generate electricity at low tide.

Ocean currents and ocean waves are caused by the winds, and so have the same disadvantage as wind when used as an energy source: they are variable and unreliable. However, it is technically possible to use ocean currents and waves to generate electricity.

29.3 Geothermal and solar energy

In mines, the temperature increases with depth, as a result of heat being conducted from Earth's core which has a temperature of about 5000 °C. In some countries, including Iceland and New Zealand, where there are hot springs, hot water from Earth's crust (geothermal energy) is used to heat buildings.

The sun's temperature is about 14 000 000 °C at the centre, and about 6000 °C at the surface. Radiant energy from the sun affects Earth in many ways. For example, the energy of wind and water currents is derived, indirectly, from solar energy (radiant energy from the sun). Winds are caused by hot air rising and cooler air taking its place; ocean currents and waves are caused by winds; and the evaporation of water makes possible the water cycle.

Energy from sunlight, stored in chemical compounds in photosynthesis, is released as heat energy when wood, coal, oil and natural gas are burned. The growth of crops other than trees, such as sugar cane, so that the sugars produced can be converted to alcohol, for use as fuel instead of petrol, is called energy farming. One disadvantage of this method of fuel production is that with world population growing, more farm land is needed for food production.

Energy from the sun is also used by solar cells, in which radiant energy is converted to electrical energy. Solar cells are expensive and they produce only small electric currents, but they can be used to recharge batteries in pocket calculators and other small electrical appliances. They are also used to provide electric power in orbiting satellites.

Solar energy can also be used directly as a source of heat energy — which is what happens naturally as the sun warms Earth. Solar panels fitted on the roofs of buildings, exposed to the sun, absorb infrared radiant energy. This heats water, and so reduces fuel bills.

Three advantages of solar energy, over fossil fuels, are that it is free, causes no pollution, and supplies cannot be exhausted. But two disadvantages are that solar energy is spread over a wide area and is difficult to collect, and it can be collected only when the sun is shining. Hot deserts — where the sun shines every day — are the best sites for collecting solar energy.

29.4 Atomic energy

Chemical bond energy is the energy stored in holding atoms together in molecules and giant structures. This bond energy is associated with the exchange or sharing of electrons (see pages 143–4). When fuels are burned this chemical bond energy is converted to heat energy.

In addition to these relatively weak forces holding atoms together, there are much stronger forces within the atom. The electrons (negative charges) are attracted by the equal number of protons (positive charges) in the nucleus. You will remember that the nucleus also contains neutrons, which have no electric charge (see page 140).

A proton has a mass 1837 times that of an electron, and the mass of a neutron is equal to that of a proton. Therefore, although electrons occupy most of the space within an atom, most of the atom's mass is concentrated in the nucleus of tightly packed protons and neutrons. These are held together by very strong forces.

The nuclear particles (protons and neutrons) are sometimes called nucleons; and nucleonics is the study of these particles. *The total number of nucleons in one atom of an element is called the atom's nucleon number or mass number*. For example, the mass number of helium (see Figure 13.2) is 4. The mass number and the atomic number can be written with the symbol for the element, as follows:

mass number: 4
symbol for helium: He
atomic number: 2

The number of neutrons in the nucleus of a particular atom can be calculated by subtracting the atomic number of the atom from its mass number. For example, an atom of aluminium ($^{27}_{13}$Al), with 13 protons in its nucleus and a mass number of 27, has $27 - 13 = 14$ neutrons.

All atoms of any one element are identical in the number of protons and electrons they contain. However, they may differ in the number of neutrons they contain. For example, a sample of chlorine gas contains both $^{35}_{17}$Cl and $^{37}_{17}$Cl atoms. Although they differ in atomic mass, these are atoms of one element. They are identical in atomic number, chemical properties, and position in the periodic table. They are called *isotopes* (Greek: *iso* = the same; *topos* = place).

Using nuclear energy

The protons and neutrons which form the nucleus of an atom are held together by such strong forces that many scientists thought it would never be possible to separate them. However, the atomic nuclei of the element uranium can be split by bombarding them with neutrons.

Naturally occurring uranium (atomic number 92) has three isotopes (U_{238} U_{235} and U_{234}). Because the isotopes of an element differ only in their mass number, it is usual when discussing isotopes to state simply the name of the element and its mass number. For example, carbon 14 is an isotope of carbon, and uranium 238 is an isotope of uranium.

When a uranium 235 atom is bombarded by neutrons, its nucleus can capture a neutron and form a very unstable nucleus which splits, forming two smaller nuclei (for example, nuclei of the elements barium and krypton). When this *nuclear fission* (splitting) occurs, energy is released together with an average of two or three neutrons.

These neutrons can collide with other uranium 235 nuclei and cause them to split — with the release of more neutrons which in turn can split more nuclei. This is an example of a chain reaction. Many millions of nuclei are split in a fraction of a second, with the release of tremendous energy. This is what happens when an atom bomb explodes.

To use atomic energy safely the chain reaction must be controlled. To control a chain reaction, some of the neutrons produced when a nucleus is a split must be absorbed so that they are not available to split other nuclei. If on average the splitting of one nucleus causes one further nucleus to split, then the reaction will continue at a steady rate. Such a controlled reaction takes place in a nuclear reactor.

In a gas-cooled nuclear reactor (see Figure 29.4) the supply of neutrons is controlled by boron steel rods. At first these control rods are in the reactor, then they are withdrawn gradually, until measuring instruments indicate that a chain reaction has started. The position of each control rod is adjusted automatically to compensate for any change in the rate of the reaction. If an emergency occurs all the rods fall into the reactor — stopping the reaction and preventing overheating.

Natural uranium sealed in magnesium alloy cans, is the fuel. It contains 139 atoms of uranium 238 for every one atom of the uranium 235 needed for the chain reaction. To increase the chance of the neutrons being captured by the uranium 235 atoms, the neutrons must be slowed down. The graphite used to do this, called the moderator, is in blocks with vertical channels for the fuel cans and control rods.

Heat energy released in nuclear fission heats carbon dioxide. This is pumped through a heat exchanger, where it heats water to produce the steam needed to rotate the alternators that generate electricity.

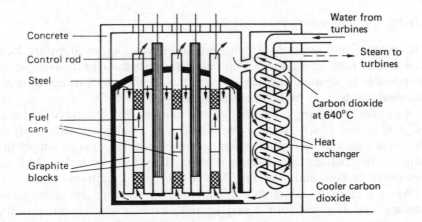

Figure 29.4 gas–cooled nuclear reactor (simplified and not to scale) (Based on Hartlepool Power Station Leaflet)

Radioactivity

More nuclear power stations would probably have been built had it not been for public concern about the disposal of radioactive waste materials from the existing nuclear power stations.

Radioactivity was discovered by a French scientist, Henri Becquerel, in 1896. He found that a photographic paper wrapped in thick black paper, in a drawer containing uranium oxide, had become fogged — as if it had been affected by light. Becquerel experimented with other uranium compounds and found they all had this effect. He concluded that they gave off invisible rays that could penetrate the thick black paper. Elements like uranium, which produce radiation, are said to be radioactive.

Because our senses are not affected by this radiation, we need to use instruments to detect it. The photographic method is too slow for most purposes because the photograph has to be developed before it can be examined. Most instruments that immediately detect radiation include a Geiger Muller tube. Radiation passing through the tube ionises the gases it contains and an electric current passes between the electrodes. The tube is connected to a loudspeaker, so that a click is heard each time an electric current passes. The more radiation is present, the more clicks are produced.

When materials are investigated with a radiation detector, all those containing elements with an atomic number greater than 83 (including uranium, thorium and radium) are radioactive. Also, some isotopes of elements with a lower atomic number are radioactive. However, most naturally occurring materials are not a danger to health.

Radioactivity is a natural phenomenon. It is affected by neither physical nor chemical changes, and is caused by changes within the atoms of an element. If a radiation detector is moved towards a radioactive source, the rate of clicking increases. As it is moved away the rate of clicking decreases. However, even if the radioactive source is removed completely the clicking does not stop. It is always possible to detect some radiation. This *background radiation* comes from naturally occurring radioactive materials on Earth, the fallout from nuclear fission tests and accidents at nuclear power stations, and from cosmic sources.

Scientists investigating radiation discovered that some rays were easily stopped, even by a thin sheet of paper, but that others would pass through a sheet of metal or a brick wall. This was not, as you might expect, because some rays were stronger than others. The explanation is that there are three different types of radiation which have been called alpha, beta and gamma radiation (from the first three letters of the Greek alphabet).

Alpha (α) radiation is a stream of particles, each of which consists of two protons and two neutrons: like the nucleus of a helium atom (see page 140). These α particles can travel only a few centimetres in air and are stopped by even a thin sheet of tissue paper.

Beta (β) radiation is a stream of electrons emitted from some unstable atoms (as neutrons in their nuclei split into protons and electrons). These beta particles can travel only about 10 cm in air and are stopped by solid objects — such as this book or a thin metal sheet.

Gamma (γ) radiation comprises rays of short wavelength, similar to x-rays. They travel long distances in air and can penetrate all except thick, dense objects such as thick lead plates.

As an atom loses particles from its nucleus it becomes a different atom with a smaller atomic mass. This is called radioactive decay (see Figure 29.5).

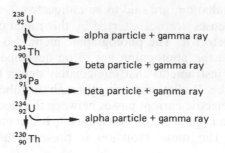

Figure 29.5 radioactive decay of protactinium (symbol Pa), and of isotopes of uranium (symbol U) and thorium (symbol Th)

Notes

Atomic number or proton number	90	91	92
Element	thorium	protactinium	uranium
Atomic mass	232.038	231	238.03

The atomic mass of some elements is not a whole number, because each of these elements is a mixture of isotopes that differ in atomic mass. The atomic mass of each isotope is a whole number

To help you understand what is happening when uranium decays, consider what happens when alpha and beta particles are lost from an atomic nucleus:

1 When a radioactive atom loses an alpha particle (2 protons + 2 neutrons), its atomic number (or proton number) decreases by two, so its mass decreases: it becomes an atom of a lighter element, lower in the series.

2 When a radioactive atom loses a beta particle (an electron), this comes from the decay of a neutron to a proton plus an electron. The nucleus retains the extra proton, so its mass hardly changes. But with an extra proton it becomes an atom of an element higher in the series.

Radiation from a radioactive substance is random. That is to say, at any time an atom of the substance may give off radiation in any direction. Because of this, it is impossible to say how long it will take for all the atoms in a sample of a radioactive substance to decay. However, scientists have calculated how long it takes — on average — for a sample of each radioactive substance to decay to half its present mass. This is called the *half life* of the substance. For some substances the half life is millions of years; for others it is only millionths of a second.

For example, radium 226 has a half life of about 1620 years. This means that if we put 1 gramme of radium 226 on one side now, in 1620 years it would have half the radioactivity and the remaining mass of radium 226 would be 0.5 grammes; and in a further 1620 years the mass of radium 226 would be 0.25 grammes; and so on.

It is about 5000 million years since Earth was formed, yet there is still some radium 226 on Earth. The reason for this is that radium is still being formed by the decay of other radioactive substances. Radium is in fact an unstable intermediate product in the breakdown of uranium to lead, which is the stable end product.

In radioactive decay, larger atoms lose matter and energy as they change to atoms of smaller mass. However, the total amount of matter and energy in the universe does not change. In 1905, nine years after Becquerel discovered radioactivity, Albert Einstein proposed his theory that matter is a concentrated form of energy; that matter and energy are interconvertible; and that from a very small mass of material a tremendous amount of energy can be obtained. This theory was proved to be correct, beyond doubt, when the first nuclear fission bomb was exploded in 1945.

The amount of energy available from the conversion of a given amount of matter is indicated by Einstein's famous mathematical equation:

$$e = mc^2$$

where e = energy in joules; m = mass in kilogrammes and c = the speed of light in metres per second.

To put this into everyday language, as much heat energy can be obtained from one gramme of material by nuclear fission as from burning 2000 tonnes of gasoline.

The law of conservation of mass (or matter, see page 89) and the law of conservation of energy (see page 105) can be combined.

The law of conservation of mass and energy can be stated: *matter and energy are interconvertible but they cannot be destroyed.*

Remember, when people speak of using up limited resources, that matter and energy are not destroyed. Matter is spread out, making it unavailable to people, because there is a limit to what can be achieved by recycling. Remember, also, that in all energy transfers some energy is wasted in the sense that it is dissipated (spread out) as heat energy — but energy is not destroyed.

30

Interdependence: organisms and their environment

The place in which an organism lives is called its *habitat*. And *the scientific study of organisms in their natural surroundings* — their environment — is called *ecology* (Greek *oikos* = a home). Ecologists study how organisms affect one another, and how they are affected by physical factors such as climate, the lie of the land, and soil conditions.

Living organisms affect and are affected by their environment, and *all the organisms in any place, interacting with one another and with their physical environment*, can be thought of as an *ecosystem*. For example, Figure 30.1 illustrates interactions between the basic components of a terrestrial ecosystem.

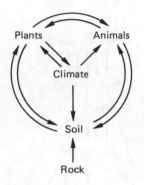

Figure 30.1 interdependent basic components of a terrestrial ecosystem. (Based on Eyres, S. R. Vegetation and Soils: a world picture, **Edward Arnold, Sevenoaks)**

Heat, cold and rain break up the soil and affect the growth of plants and animals. These influences are represented in the diagram by the three arrows pointing from the word climate. Other influences are also represented by arrows. The cover of vegetation has a moderating effect on the temperature and water content of the air near the ground. The roots of plants help to break up the soil, and the bodies of dead plants and animals decay and become part of the soil. Chemicals in the soil water are dissolved from the rock particles and from decaying organic matter. The growing plants absorb some of these chemicals, and provide a variety of places in which animals live. Many animals affect both the growth of individual plants and the development of plant communities.

Even a handful of a natural undisturbed soil is teaming with living organisms and can be thought of as an ecosystem. And people have begun to realise that all life on Earth is part of one global ecosystem. All organisms are interdependent, people could not live alone, and the things people do in one part of the world can affect the lives of people and of other organisms in other parts. Words such as habitat, ecology and ecosystem are becoming part of everyone's vocabulary, so it is important to know what they mean.

30.1 Producers and consumers

Organisms with chlorophyll produce their own food and are therefore called producers. All animals are consumers: herbivores (feeding only on plants), carnivores (feeding only on animals), or omnivores (feeding on both plants and animals). Some nutritional relationships can be represented as a *food chain*, for example:

green plants ⟶ rabbits ⟶ people.

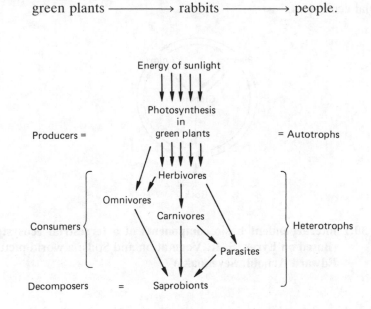

Figure 30.2 flow of matter and energy through organisms. (From Barrass, R. Human Biology Made Simple)

The sun is the only source of energy available to living organisms. The arrows in Figure 30.2 represent the flow of energy through different kinds of organisms. This looks more complicated than a food chain but it is still an oversimplification of the interactions involved in feeding and *energy flow*. A *food web* (Figure 30.3) is an attempt to represent the organisms in a community that are consumed and the organisms which consume them.

In Figure 30.4 producers and consumers are represented at different levels in a *pyramid of numbers*. The producers in any place support a certain number of primary consumers, which provide food for fewer secondary consumers, and so on. Note also that the *mass of living matter* (the *biomass*) decreases: a lot of vegetation supports a large number of primary consumers (with a total mass much less than that of the vegetation) and there are fewer secondary consumers (with a total mass less than that of the primary consumers), etc.

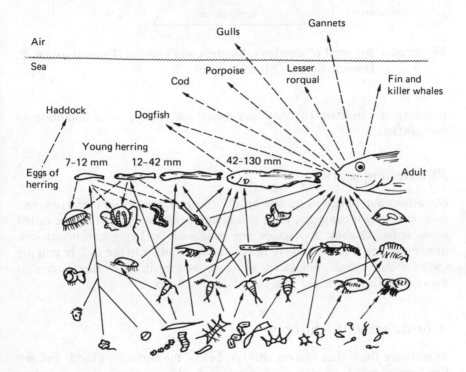

Figure 30.3 A food web. Each arrow means eaten by. **The complete and broken lines indicate the place of the herring** *Clupea harengus* **in the food web. (From Barrass, R.** Modern Biology Made Simple)

Biomass decreases at successive levels in the pyramid because some food is not digested by the animals and some of the absorbed food is used as an energy source. There is a loss of energy as heat.

One result of the nutritional interdependence of all organisms in any place is that the number of individuals of each species may remain fairly constant from year to year. There is a balance of nature (an ecological

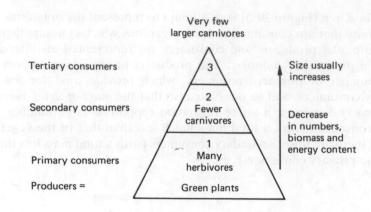

Figure 30.4 **pyramid of numbers, biomass and energy. (From Barrass, R.** Human Biology Made Simple)

balance): a condition of constancy based on change — a condition of homeostasis.

30.2 Decomposers

Organisms that obtain nutriment by absorbing organic molecules from their environment (for example, yeast absorbing sugars: see page 173) are called *saprobionts*. Organic molecules are produced by living organisms, are present in their bodies, and are released after death into the soil. In making use of these organic molecules, saprobionts contribute to the process of decay and, therefore, are also called decomposers.

A bread mould: Rhizopus

Moulds are fungi that live on and just below the surface of food, and are first conspicuous as disc-shaped patches. *Rhizopus* forms grey-white patches.

Microscopic spores of *Rhizopus* are carried in the air, and they germinate if deposited on stored fruit, damp grain or bread. A tubular outgrowth called a hypha penetrates the bread, secreting enzymes that digest the food and absorbing the products of digestion. The hypha grows at its tip, branches and becomes a mass of threads called a mycelium. At the surface of the bread some erect hyphae project into the air (Figure 30.5). Each of these unbranched aerial hyphae is called a sporangiophore because a spore case or sporangium develops at its tip.

All the hyphae have a wall outside the surface membrane and a large central vacuole, and many nuclei share the same cytoplasm. That is to say,

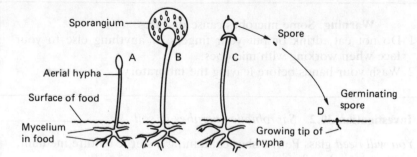

Figure 30.5 life cycle of Rhizopus, **a fungus: (A) developing sporangiophore; (B) sporangium containing spores; (C) spores released; (D) a spore germinating on a suitable food develops into a new mycelium. (From Barrass, R.** Modern Biology Made Simple)

the hyphae are not divided into cells. Streaming of cytoplasm carries food molecules, absorbed at the tips of the hyphae, to all parts of the mycelium.

A cross wall forms in each sporangiophore, between the sporangium and its stalk. Then each nucleus in the sporangium, with some cytoplasm, forms a separate spore with a spore wall. So an organism which developed from a spore produces many spores. These are released when the spore case breaks open and may be carried in the wind.

Like any other organism, *Rhizopus* differs in size and shape at the successive stages in its life cycle. Its form at each stage is adapted to its place in the life cycle: the mycelium is concerned with nutrition and growth; the sporangiophores with reproduction; and the spores with dispersal and the survival of the species.

Investigation 30.1 *Saprobiotic fungi on damp bread*

You will need a slice of bread and a plastic bag.

Proceed as follows
Expose the bread to air for a few hours then enclose it in the plastic bag. Examine the bread once a week, **without opening the bag,** and record your observations. You will see the different colours of spores produced by different kinds of fungi. Dispose of the sealed bag.

> ⚠
>
> **Warning**. Some microbes cause diseases.
> 1 Do not eat, drink or put your fingers or anything else to your face when working with microbes.
> 2 Wash your hands before leaving the laboratory.

Investigation 30.2 *Saprobiotic microbes in soil*

You will need glass Petri dishes containing a sterile culture medium, and some fresh soil.

Proceed as follows
1 Prepare the culture medium. Dissolve 1 g beef extract, 0.2 g yeast extract, 1 g peptone and 0.5 g sodium chloride in 100 cm³ distilled water in a conical flask. Stir in 1.5 g agar. Plug the flask with non-absorbent cotton wool.
2 To ensure there are no organisms in the culture medium at the start of your investigation, sterilise the medium and the Petri dishes in a pressure cooker (15 min at 1 kg per cubic centimetre).
3 Close all windows and doors and work where there are no draughts. Allow the medium to cool to about 55 °C for pouring. Place the petri dishes on a flat surface. Raise the lid slowly at one side so there is enough space to pour in some of the medium. Pour in enough to cover the bottom of the dish. Lower the lid slowly. Prepare as many dishes as you require (one for each soil sample).
4 Mix some fresh soil with an equal volume of sterile distilled water. Pour this mixture on to a plate of culture medium. Leave for 30 s then pour away the soil and water. Replace the lid on the dish. Label the dish. Fix the top of the dish to the bottom with tape diagonally accross the dish.
5 Incubate the plate upside down at 25 °C or at room temperature. Bacteria and fungi from the soil will multiply. You cannot see individual bacteria but you will see patches of large numbers of bacteria and patches resulting from the growth of mould fungi.
6 At the end of Investigation 30.2, examine the Petri dish but do not remove the tape securing the lid to the base. Place the unopened dish in disinfectant (fresh 40 per cent formaldehyde).

30.3   The circulation of elements in nature

The carbon cycle

Carbon, from carbon dioxide in air, is assimilated by green plants in photosynthesis (see page 163). The carbon then becomes part of the organic molecules of the green plant. Animals absorb carbon-containing molecules after digesting either plants or other animals that have fed on plants. When plants and animals die, saprobionts contribute to their decay by digesting and absorbing organic molecules. In the respiration of all organisms, carbon-containing molecules are used as an energy source and carbon dioxide is returned to the air (see page 173). This circulation of carbon in nature, between organisms and their environment, can be represented as a carbon cycle (see Figure 30.6). Note also that in burning timber and fossil fuels (see page 109) carbon-containing molecules are completely oxidised and carbon dioxide is returned to the air.

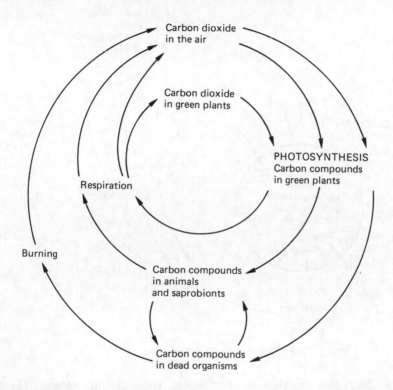

Figure 30.6 the carbon cycle. (From Barrass, R. Modern Biology Made Simple)

Mineral element cycles

Plants absorb mineral ions. Some are incorporated in organic molecules and others remain in solution in the cell fluids. Animals absorb these ions from their food. When plants and animals die, and when undigested food is egested by animals, these ions are available to saprobionts. The excretions of animals also include mineral ions which may be absorbed by green plants or by saprobionts.

The only source of mineral ions is by solution of rock particles in the soil, but once absorbed by living organisms the supply of mineral ions in any place depends upon their circulation from one organism to another, from organisms to their environment, and from the environment to other organisms (see Figure 30.7).

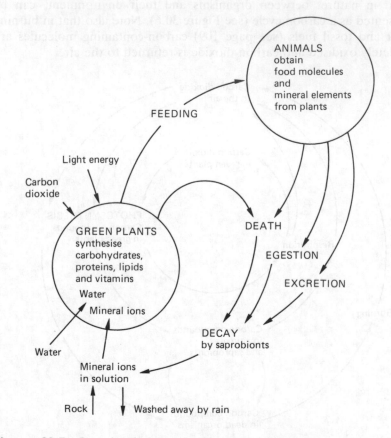

Figure 30.7 the circulation of mineral elements in nature. (From Barrass, R. Modern Biology Made Simple)

The nitrogen cycle

Amino acid molecules, from which proteins are formed, contain nitrogen — which is therefore an essential element in living organisms. Nitrogen is not present in rocks but it does occur in air (see page 115) and as a result of the activities of living organisms it is present in the soil.

Plants absorb nitrogen-containing ions from the soil and use them in the synthesis of amino acids. Animals obtain amino acids from their food —

either directly or indirectly from plants. The amino acids from dead organisms are absorbed by saprobionts.

Ammonium ions are added to the soil when dead organisms decay and also from the nitrogenous excretions of animals. Some saprobiotic bacteria in the soil, called *nitrifying bacteria*, obtain energy by converting ammonium ions to nitrite ions and to nitrate ions (see page 148).

Other soil bacteria are called denitrifying bacteria because they break down nitrate ions. As a result of their activities, nitrogen gas is returned to the air. The amount of nitrogen available to living organisms is reduced by this *denitrification*. However, there are other soil bacteria which absorb nitrogen gas from the soil atmosphere and use it in producing amino acids. When these *nitrogen–fixing* bacteria die and decay, nitrogen–containing compounds are added to the soil. Other nitrogen–fixing bacteria live in swellings called root nodules which are formed by leguminous plants (for example, beans and other plants with pods).

All these aspects of the circulation of nitrogen in nature are shown in Figure 30.8. In addition, nitrogen and oxygen combine in the atmosphere during lightning discharges, forming nitrous and nitric oxides which dissolve in rain and so are added to the soil.

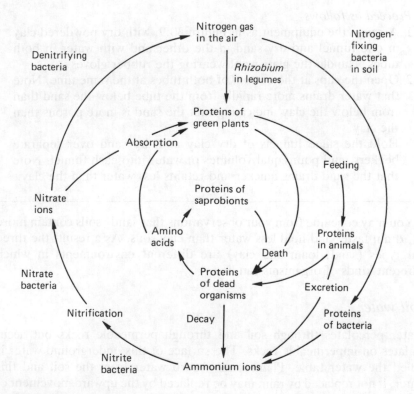

Figure 30.8 the nitrogen cycle. (From Barrass, R. Modern Biology Made Simple)

30.4 The soil as an ecosystem

Soil is a mixture of (1) small rock particles, (2) water, (3) chemicals in solution, (4) air spaces, (5) living organisms, and (6) humus — the decaying remains of organisms. The properties of a soil are determined by the kind (or kinds) of rock from which it was formed, and by the size of the rock particles present in the soil.

Small rock particles

Soils in which the small particles are mostly above 0.02 mm diameter are called sandy soils. Soils with much smaller rock particles (mostly less than 0.002 mm diameter) are called clay soils. Loam soils have most particles of intermediate size mixed with some sand and clay.

Investigation 30.3   *Some properties of soils*

You will need the equipment labelled in Figure 30.9, some dry clay, some dry sand and rubber gloves. The funnels and tubes must be identical.

Proceed as follows
1 Arrange the equipment as in Figure 30.9, with dry powdered clay in one funnel and dry sand in the other and with water in both tubes. Handle the glass wool wearing the rubber gloves.
2 Open the clips at the bottom of both tubes at the same time. Note that water drains more rapidly from the tube below the sand than from below the clay, indicating that the sand is more porous than the clay.
3 Hold the same funnels of dry clay and dry sand over separate beakers, and pour equal volumes of water into each funnel. Note that the sand drains quicker and retains less water than the clay.

You may conclude from your observations that sandy soils contain more air, drain faster and hold less water than clay soils. As a result, the three soil types (sand, loam and clay) are different environments in which different kinds of organisms can live.

Soil water

Water percolates through soil and through permeable rocks but accumulates on impermeable rocks. The surface of this underground water is called the water table. Plant roots absorb water from the soil and this water, if not replaced by rain, may be replaced by the upward movement of water in the soil by capillarity (See page 42).

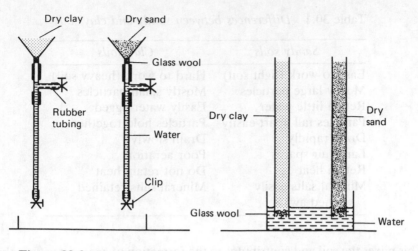

Figure 30.9 porosity and permeability of soils

Figure 30.10 water rising through clay or sand by capillarity

---

Investigation 30.4 *Water rising through soil*

You will need two glass tubes, a glass dish, rubber gloves and the other materials illustrated in Figure 30.10.

Proceed as follows
1 Pack one end of each tube with glass wool, wearing rubber gloves.
2 Fill one tube with dry sand and the other with dry powdered clay.
3 Place the tubes in the dish, in water, as in Figure 30.10. Note in which tube the water rises: (a) faster, and (b) further.

The amount of water in a soil depends partly on rainfall and the lie of the land, but also on the water-retaining capacity of the soil — due partly to the humus content, because humus acts like a sponge, but also to capillary forces between the small rock particles. Clay helps to retain water in soil. Capillary forces also cause the small particles of rock and humus to hold together.

Living organisms in soil

Earthworms feed on soil and digest organic matter. By burrowing, they contribute to soil aeration and drainage. They pull vegetable matter into their burrows and deposit worm casts in the soil and on the soil surface.

Table 30.1   *Differences between sandy and clay soils*

Sandy soils	Clay soils
Easy to work (light soil)	Hard to work (heavy soil)
Mostly large particles	Mostly small particles
Retain little water	Easily waterlogged
Particles fall apart easily	Particles hold together
Drain rapidly	Drain slowly
Large air spaces	Poor aeration
Retain heat	Do not retain heat
Mineral salts easily washed away	Mineral salts retained

This mixes the soil and contributes to the formation of aggregates (crumbs) of organic and inorganic matter.

Soil also contains a great variety of other organisms, including millions of microbes in each cubic centimetre of fertile soil. Many bacteria, fungi, protists and nematode worms are saprobiotic. They contribute to decay and help to retain nutrients in the soil (see Figure 30.6, Figure 30.7 and Figure 30.8).

31 Hygiene in the home and community

31.1 Parasites of people

When your body is not working properly you feel ill — you are suffering from a disease. Some diseases are the result of inherited defects and some are due to dietary deficiencies. But this chapter is about diseases caused by living organisms, called parasites, that may invade your body.

Parasites benefit from this close association — obtaining shelter and food. But the organism invaded by a parasite (called its host) does not benefit. Indeed the host may be harmed by the activities of the parasite, which is then called a pathogen (Greek: *pathos* = disease; *genesis* = producing).

Flatworms that cause disease

Blood flukes. The blood fluke *Schistosoma* lives in the blood of people. Its body is flat, like a narrow leaf about 10 mm long. It has two suckers by which it maintains its position in a vein. The fluke's eggs pass through the wall of its host's intestine or urinary bladder and leave the body with the faeces or urine. If the eggs end up in water, larvae hatch and may enter the body of a water snail. This mollusc is the intermediate host in which the blood fluke reproduces asexually (without the formation and fusion of gametes), producing many offspring. When these leave the snail they penetrate the skin of anyone who bathes or wades in the water.

Blood flukes cause a debilitating disease called schistosomiasis (or bilharzia) which affects millions of people and causes untold misery in many countries in the Far East, Middle East, Africa and South America. Control of the disease depends on killing the snails. But people can help themselves by ensuring that their faeces and urine are disposed of properly. Remember, also, that it is not safe to bathe in natural waters in any country where some people suffer from schistosomiasis.

Tapeworms. The intestines of animals with backbones — including pets, farm animals and people — are the habitats of tapeworms. Each tapeworm has a knob-like scolex (about 1 mm diameter). This has suckers and is embedded between the villi of its host's intestine (see Figure 19.2). The scolex has a cylindrical part that extends into a flat tape made up of many pieces called proglottids (see Figure 31.1). New proglottids are formed by the scolex, and once formed they grow and develop.

The proglottids have no digestive system but they absorb food molecules resulting from the host's digestion of food. As a result the host may be weakened by the presence of the parasite.

Each proglottid produces first sperms and then eggs: it is neither male nor female but hermaphrodite. Older proglottids contain fertilised eggs which develop into embryos. The largest and oldest proglottids, packed with embryos, break away from the end of the tape and pass out with the host's faeces. An embryo cannot then develop further unless it enters another vertebrate host (see Figure 31.1), called the intermediate host, in which it develops into a bladderworm in a muscle. For example:

	Tapeworm	**Bladderworm**
Beef tapeworm	in people	in cattle
Pork tapeworm	in people	in pigs

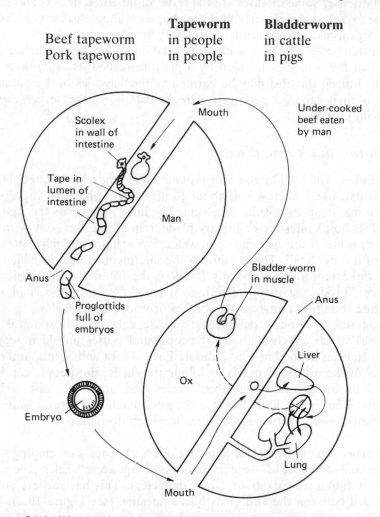

Figure 31.1 life cycle of the beef tapeworm *Taenia saginata.* **(From Barrass, R.** Human Biology Made Simple)

Control of tapeworms depends on effective sanitation and sewage disposal, which breaks the link between people and the intermediate hosts. However, people can get tapeworms by eating insufficiently cooked beef or pork that contains bladderworms. So never eat insufficiently cooked meat or meat products (such as sausages).

Roundworms that cause disease

Ascariasis. *Ascaris* has a cylindrical body 20 to 25 cm long that tapers at both ends. It lives in the intestine of its host, ingesting food digested by the host and so reducing the amount of food available to the host. Thousands of eggs pass out with the faeces and may contaminate crop plants — especially if faeces are used as manure. The eggs can live for several years in the soil and people may ingest them (be infected) if there is soil in their fingernails or if they eat contaminated food.

Threadworms. *Oxyuris*, the threadworm (about 1 cm long), lives and reproduces in the large intestine, causing irritation around the anus. A person's fingernails may be contaminated by eggs — which can be transferred directly to the mouth (especially if children suck their thumbs or bite their nails). This introduces more worms into the gut and the condition persists.

Hookworms. *Ancylostoma*, the hookworm (about 1 cm long), lives in the small intestine of its host (see Figure 31.2). The adult sucks blood and causes anaemia. The debilitating effect of a heavy infection reduces a person's ability to withstand other diseases. Hookworms undermine the health of people throughout the tropics and subtropics. Their spread can be prevented by the proper disposal of faeces and by effective sewage treatment. Because hookworms can bore through the skin (see Figure 31.2), people should wear shoes and they should neither pass faeces where other people may walk nor use infected faeces as manure.

Filariasis (elephantiasis). *Wuchereria*, the filaria worm, lives in lymphatic vessels (see page 224). The adults are quite large (female = 10 cm; male = 4 cm) and they may block the lymph vessels causing the tissues to swell — hence the name elephantiasis. The larvae, however, are only 0.3 mm long and they pass from the lymph vessels into the blood stream.

In 1878 Patrick Manson discovered these larvae in mosquitoes that had bitten infected people. This was the first demonstration that a blood parasite of people could also live in an insect. Manson did not know how the parasite was passed from the infected mosquitoes to other people. We now know that the mosquito's mouthparts become contaminated with filaria larvae which enter the human body through the wound made when the mosquito bites another person. In this way the mosquito is essential for the spread of this disease from person to person, and control of the disease

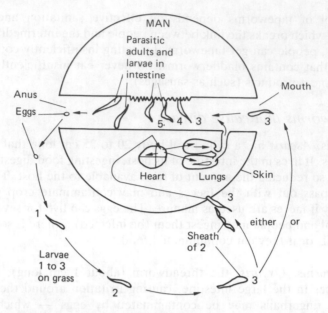

Figure 31.2 life cycle of the hookworm *Ancylostoma duodenale.* **Numbers 1–5 refer to stages of development, 1–3 being larval stages. (From Barrass, R.** Human Biology Made Simple)

depends on the control of mosquitoes (see below) in all tropical and subtropical countries.

Protists that cause disease

The malarial parasite. *Plasmodium*, the malarial parasite, is a protist (see page 72) that lives in the liver epithelial cells and in the erythrocytes of people who have malaria. It absorbs nutriment and destroys liver cells and erythrocytes.

Malaria is caused by the regular release of the parasite's waste products into the blood of its host, every 24 hours or every 48 hours. These cause the periodic fevers characteristic of the different forms of malaria. More people die of malaria than from any other cause.

Female *Anopheles* mosquitoes feed on blood. They ingest *Plasmodium* when they suck blood from someone who has malaria (Figure 31.3A). When they next feed, the mosquitoes may inject parasites into the blood of another person. This is likely to happen because, before they feed, mosquitoes inject saliva into the wound. The saliva contains an anticoagulant, which prevents the blood from clotting, and this enables the mosquito to suck up the blood. But if the mosquito is infected with *Plasmodium*, these parasites will be injected with the saliva (Figure 31.3B). This is the only way in which malarial parasites can pass from one person to another.

Quinine and many synthetic drugs have been used to treat malaria (see page 375). However, if contact between people and mosquitoes can be reduced there is less chance of people being infected. In places where there is malaria, therefore, people should sleep under mosquito nets. Residual insecticides are also used, on the inside walls of houses, in an attempt to kill those mosquitoes that are most likely to bite people.

The eggs, larvae and pupae of mosquitoes live in water (see Figure 31.4) and mosquito control operations usually include: (1) draining standing waters to reduce the number of mosquito breeding sites, and (2) introducing minnows (mosquito–eating fish) into ditches and wells as a method of biological control.

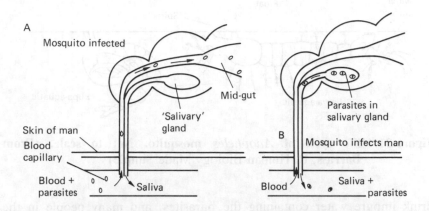

Figure 31.3 mosquito feeding on a man or woman: (A) ingests malarial parasites; and (B) transmits malaria when it feeds again. (From Barrass, R. Human Biology Made Simple)

Other parasitic protists. *Trypanosoma* causes sleeping sickness of people and a disease of horses and cattle called nagana. These diseases are confined to Africa, south of the Sahara desert, and *Glossina*, the tsetse fly, is the intermediate host. Tsetse flies feed on blood and transmit trypanosomes from one mammal to another. They require shade and blood meals, and they do not live in towns or in cultivated areas where most of the natural vegetation has been cleared. In other places, the people of Africa should recognise the value of wild game animals which do not suffer from nagana. They make more efficient use of the varied bush and grassland vegetation than do introduced cattle. They should be conserved and cropped as a source of food as well as being a source of revenue from tourism.

Most species of *Amoeba* live in fresh water (see page 72) or in moist soil but some live inside other organisms. Amoebic dysentery (in which people pass blood and suffer from severe diarrhoea) is caused by *Entamoeba histolytica* which lives in the large intestine. People can be infected if they

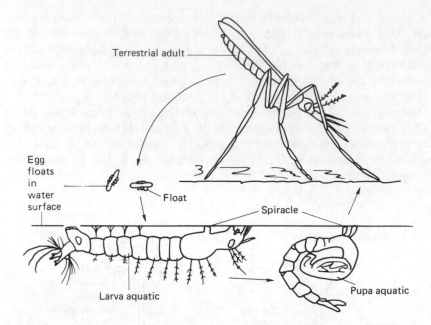

Figure 31.4 life cycle of *Anopheles* **mosquito. Not to scale. (From Barrass, R.** Human Biology Made Simple)

drink impure water containing the parasites, and many people in the tropics suffer from this disease.

Investigation 31.1   *Protists that cause disease*

Examine prepared microscope slides or colour transparencies of *Plasmodium* and *Trypanosoma* in blood smears.

Investigation 31.2   *Stages in the life cycle of a mosquito*

Examine prepared microscope slides or colour transparencies of stages in the life cycle of an *Anopheles* mosquito (compare with Figure 31.4).

Fungi that cause disease

Many fungi cause diseases of plants (see page 22) and some cause diseases of animals. **Ringworm**, for example, is not caused by a worm. It is a fungus disease of the skin and nails. You can catch it from a pet cat or dog, or if you try on a hat in a shop, or if you use someone else's pillow, brush or comb.

Thrush, another fungus disease, is caused by a yeast that grows and reproduces in the mucus of the mouth cavity and throat, forming a creamy–white layer.

Bacteria that cause diseases

Parasitic bacteria. In 1866 Louis Pasteur investigated a disease of silkworm moth caterpillars, reared in France for silk production, and he was the first to demonstrate that bacteria could cause a disease of animals.

In 1876 Robert Koch, in Germany, (1) isolated bacteria from sheep suffering from anthrax, (2) grew these bacteria in pure culture in the laboratory, and (3) injected bacteria from the pure culture into healthy sheep which then developed anthrax.

We now know that many diseases of people are caused by parasitic bacteria (see Table 31.1). Some enter the body through the air passages

Table 31.1 *Diseases of people caused by bacteria*

Disease	*Method of spread of disease*
Airborne diseases	
Diphtheria	By breathing in contaminated air, especially
Whooping cough	the very fine droplets spread when people
Scarlet fever	cough and sneeze without covering their
Tuberculosis	mouths with a handkerchief.
Tonsilitis	
Pneumonia	
Leprosy	
Water-borne and food-borne diseases	
Cholera	Contaminated water; contaminated hands,
Dysentery (bacillary)	taps and door handles; house-flies; poor
Typhoid fever	hygiene; contaminated baby-feeding bottles
Gastro-enteritis	and contaminated food.
Food poisoning	
Diseases introduced through wounds	
Bubonic plague	Bite of rat fleas.
Relapsing fever	Contamination of wounds by lice.
Tetanus	Contamination of wounds with dirt.
Leprosy	Contaminated scratches.
Contact diseases	
Syphilis	Sexual intercourse, and, rarely, by kissing
Gonorrhoea	

and lungs (airborne or *infectious diseases*), others through the alimentary canal (*water or food-borne diseases*), and others through the skin (contact or *contagious diseases*). Bacteria may also enter the body directly through wounds.

The fact that people can transfer a disease from a patient who is suffering from the disease to a previously healthy person was demonstrated by Semmelweiss in 1846. By insisting that doctors and nurses washed and rinsed their hands in chlorinated lime solution, he reduced the death rate from a fever in the maternity hospital in Vienna from 7 per cent to 1.29 per cent without knowing the cause of the disease.

The chlorinated lime solution was what we now call a *disinfectant: a chemical used to kill harmful bacteria* and so to reduce the chances of infection. In 1876 Joseph Lister started *antiseptic surgery* in Britain. He used a carbolic acid (phenol) solution to *sterilise* both his instruments and the wounds caused by surgery. We now know that this disinfectant reduced the chances of bacteria contaminating the wounds.

A wound that is contaminated may become septic. White blood cells, feeding on the bacteria, form the white pus which is an indication of *sepsis*. To prevent sepsis, the skin should be kept clean and any wounds or scratches, however small, should be cleaned at once and treated with an *antiseptic* (a chemical used to prevent sepsis) and covered with a clean dressing.

An antiseptic is a disinfectant that is mild enough for use in an open wound to prevent the multiplication of bacteria without damaging the body. However, because disinfectants and antiseptics are poisonous they must be kept away from children and they should be properly labelled.

In many hospitals nowadays, antiseptics are not used in surgery because they can interfere with wound healing. Instead contamination is avoided by cleaning the skin before the operation and by ensuring that all people in the operating theatre wear sterilised clothing, gloves and face masks. Sterile instruments and dressings are used. This is caused *aseptic surgery*.

Preventing disease caused by parasitic bacteria. Parasitic bacteria absorb food molecules from the host's cells and body fluids, and pass waste products into the host's tissues. Some of these waste chemicals are toxic (poisonous) and they are called *toxins*.It is these toxins that cause diseases; and the host's responses to the presence of toxins are the signs or symptoms of the disease caused by bacteria of a particular kind. The time between catching a disease (when the bacteria enter the body) and the appearance of symptoms is called the incubation period. In the presence of toxins the host's white blood cells produce *anti-toxins*, which neutralise the toxins. Even in the absence of treatment this may result in the host's recovery from the disease. After recovery, as a result of the presence of anti-toxins in the blood and other body fluids, the host may not suffer from the same disease again: we say that it is *immune* to this disease.

Pasteur found that anthrax bacteria were weakened when cultured at a higher than normal temperature. In 1881 he demonstrated dramatically that when the weakened bacteria were injected into sheep, a technique called inoculation, the sheep did not suffer from anthrax. More than this, these sheep recovered from the disease when, in a later experiment, they were infected with normal (unweakened) anthrax bacteria. Pasteur concluded that the first inoculation, with weakened bacteria, had given these sheep an *immunity* to anthrax.

The technique of *immunisation* has been developed and healthy people can now be treated so they develop immunity to many of the diseases caused by bacteria — and so do not suffer from these diseases. As a result many lives have been saved. This is an example of preventive medicine, and prevention of disease is better than cure.

Curing diseases caused by parasitic bacteria. The use of chemicals against parasites in the body of their host is called *chemotherapy*.

In 1928 Alexander Fleming, a British microbiologist, observed the mould fungus *Penicillium* growing as a contaminant in a culture of bacteria. He noticed that there was a clear area around the fungus where no bacteria were growing. He concluded from this observation that the fungus must be excreting something that inhibited the growth of bacteria.

Later workers extracted from the fungus a chemical that was toxic to bacteria. This drug, called penicillin, was the first of the *antibiotics* (Greek: *anti* = against; *bios* = life). It was introduced in the 1940s. Since then, other antibiotics have been discovered. They are all used in the treatment of diseases caused by bacteria — for which previously there was no cure.

Viruses that cause disease

Viruses (0.01 to 0.3 μm diameter) are the smallest organisms. Each kind of virus grows and reproduces only in the cells of an appropriate host, and different kinds of virus are transmitted from host to host in different ways (see Table 31.2).

Edward Jenner, a British country doctor and naturalist, observed that people who had suffered from cowpox did not catch smallpox. In 1789, nearly 100 years before Pasteur's work on anthrax (see above), Jenner produced an immunity to smallpox by taking matter from a cowpox pustule and introducing this into people through small punctures in their skin. Jenner did not know what caused the disease; nor did he understand the reason for the success of this treatment (called *vaccination* – Latin: *vacca* = a cow). Indeed, no one knew of the existence of viruses until 1898.

As a result of the use of vaccination to prevent smallpox, this disease has been eradicated. The last case was reported in 1978. Other effective vaccines have been developed for the prevention of other virus diseases — including measles, rubella (German measles) and poliomyelitis. As a result many lives have been saved.

Antibiotics do not kill viruses but they may be given to people suffering from virus diseases — to prevent secondary bacterial infections.

Table 31.2   *Diseases of people caused by viruses*

Disease	Method of spread of disease
Airborne diseases	
Common cold	Breathing in contaminated air, especially the
Influenza	very fine droplets spread when people cough
Measles	and sneeze without covering their mouths
Rubella (German measles)	with a handkerchief.
Chicken-pox	
Mumps	
Water-borne and food-borne diseases	
Poliomyelitis	From contaminated food or water or by droplet infection.
Infectious hepatitis	From infected food handlers; by contact and droplet infection.
Diseases introduced through wounds	
Yellow fever	Bite of *Aedes* mosquito.
Rabies	Bite of a rabid dog.
Serum hepatitis	From contaminated instruments, especially the hypodermic needles used by drug addicts (see page 231).
Contact disease	
AIDS	Sexual intercourse, contaminated hypodermic needles (see page 253).

31.2   Cleanliness in the home and community

Medicine is concerned with maintaining a healthy body, and hygiene with personal cleanliness and maintaining a clean environment in an attempt to prevent the spread of diseases caused by parasites.

Clean food

The basic requirements for food hygiene are as follows:
1 Wash your hands after going to the lavatory and before eating or handling food.
2 Clean with soap and water all surfaces, containers and utensils used in food preparation, display and storage — and keep them clean.
3 Cough into a clean handkerchief, not into your hand, into the air, or on to food; and do not touch your mouth or any other part of your body when preparing food.
4 Cover food to exclude dogs, cats, rats, mice, cockroaches and houseflies.
5 Eat fresh food or food that has been properly preserved.

Food poisoning may be caused by *Salmonella*, a saprobiotic bacterium that occurs in polluted water, faeces and raw meat. Cooking destroys *Salmonella* but it does not destroy the toxins it produces (see page 356). It is especially dangerous: (1) to keep cooked meat near uncooked meat or to use the same serving utensils (or the hands) for both, (2) to reheat meat dishes, (3) to cook frozen meat that has not been properly defrosted, or (4) to re-freeze frozen food once it has been allowed to thaw.

Food poisoning may also be caused by the soil bacterium *Clostridium*. Inadequately processed home-preserved foods are the most usual cause of trouble. Both the bacteria and the toxins they produce are destroyed by boiling food for 15 to 20 minutes.

Sanitation

The control of some diseases depends on the proper disposal of faeces and urine (see tapeworms, page 351; blood flukes, page 349; ascariasis, page 352; threadworms, page 351; hookworms, page 351; amoebic dysentery, page 353; and water-borne diseases caused by bacteria and viruses (see Table 31.1 and Table 31.2).

Faeces, therefore, should never be left exposed on the ground. They should at least be buried. Where there are houses, sanitation — maintaining clean conditions in and near buildings — depends on other means of disposal. Conservation methods involve the construction of different types of latrines: the faeces and urine being either removed regularly or allowed to decompose. Water-carriage disposal is possible only in countries where: (1) there is enough water to flush urine and faeces from sanitary fittings, (2) the houses have a piped water supply, sanitary fittings and drains, and (3) there are properly constructed sewers and treatment works.

In one method of treating sewage, solids are allowed to settle in a sedimentation tank (see Figure 31.5). The fluid effluent from the sedimentation tank (containing inorganic ions and organic molecules in solution, and small organic particles in suspension) is spread on a filter bed from a rotating sprinkler. Saprobiotic bacteria and fungi, protists, small worms

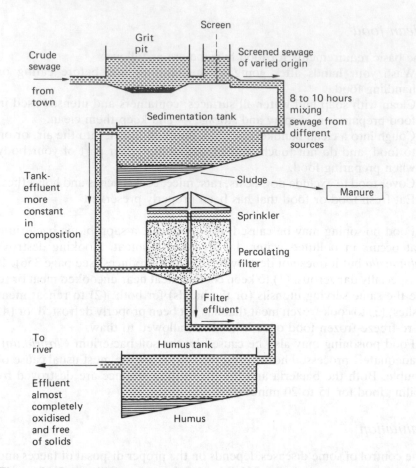

Figure 31.5 the course of sewage through a treatment plant which includes a percolating filter. (From Barrass, R. Human Biology Made Simple)

and insect larvae live in the filter bed on the surface of small stones. They absorb nutrients from the liquid, feed on the small organic particles, or feed on each other, as part of a complex food web. As a result, the water that flows from a sewage treatment works into a river should contain very little organic matter. In some countries this effluent is so clean that water from rivers, below sewage works, can be extracted and purified (by further filtration and chlorination) for re-use (recycling).

Disposing of refuse (garbage)

Refuse bins. Effective sanitation and refuse disposal reduce unpleasant odours in and near buildings. These odours attract houseflies. All insects that walk on rotten food or faeces are a danger to health because they may

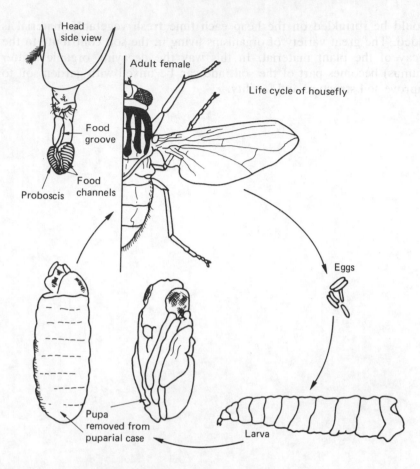

Head side view
Food groove
Food channels
Proboscis
Adult female
Life cycle of housefly
Eggs
Larva
Pupa removed from puparial case

Figure 31.6 life cycle of housefly Musca domestica. **Houseflies lay eggs on horse manure and waste food, and the larvae feed on this decaying organic matter. (Based on Barrass, R. Rearing houseflies and their use in laboratory practical work.** Journal of Biological Education, **10,** 164–8)

later walk on food in your kitchen. Microbes that cause disease may be carried on their feet and mouthparts (see Figure 31.6) and in their faeces. Therefore, all food should be kept covered and any waste food should be placed at once in a closed container before its safe disposal. Refuse bins should be strong enough to keep out birds, rats and mice, as well as flies.

Refuse disposal. Methods by which people can dispose of household waste materials include: (1) boiling fresh waste food and then feeding it to pigs or poultry, (2) burning or burying it, and (3) composting.

All waste organic material from a garden, such as parts of plants not required for food by people, can be placed on a compost heap. A little soil should be sprinkled on the heap each time fresh vegetable material is added. The great variety of organisms living in the soil contribute to the decay of the plant material. In this way the decaying organic matter (humus) becomes part of the soil, and can be mixed with garden soil to improve soil structure and fertility.

32 The electronics revolution

Methods of communication, for example by letter and by telephone, are sometimes called communication systems. They make information transfer possible between people who can neither hear nor see each other directly. In a letter, or in a recorded telephone conversation, information is also stored.

In any communication system there is an input, a signal that may be carried a long distance, and an output (see Figure 32.1).

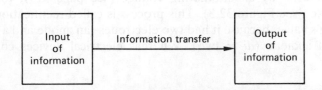

Figure 32.1 essential parts of any communication system

The development of modern methods of communication (including radio and television), and of modern methods of handling and storing information (for example, using word processors and computers), has been based on discoveries concerning the behaviour of electrons.

In 1820 Oersted noticed that an electric current in a conductor produced a magnetic field (see page 302). Then, in 1864, James Clerk Maxwell, a Scottish physicist, found that a changing electric field induced a changing magnetic field, which in turn induced a changing electric field, and so on. He called this radiation from a source *electromagnetic radiation* (see Table 32.4).

In 1888 Heinrich Hertz, a German physicist, produced and detected the electromagnetic waves (see Table 32.4) we now call radio waves, which at first were called hertzian waves. Other scientists sent messages short distances, from a transmitter (which produced radio waves) to a receiver (which received radio waves), by switching on the transmitter for shorter or longer periods to represent the dots and dashes of morse code. Then in 1901 Guglielmo Marconi, an Italian, earthed both the transmitter and the receiver (see page 321) and provided each of them with an aerial. As a result, he was able to send the first message by radio across the Atlantic Ocean. Wires and cables, linking places, were no longer essential for rapid international communication. At once, the world seemed a much smaller place.

32.1 Electronic devices

In 1883 Thomas Edison, an American, perhaps best known as the inventor of the phonograph and motion-picture projector, made an electric light bulb that had a fine wire screen near the heater filament. Whilst trying out this bulb he discovered, by accident, that when the wire screen was made positive with respect to the heater filament, electricity flowed in the circuit. Electrons were crossing the gap between the filament and the screen.

In 1904 John Fleming inserted a metal plate, instead of the fine wire, near the heater filament (see Figure 32.2B). When this plate was positively charged it attracted electrons (acted as an anode, see page 155), and so completed the circuit. But when negatively charged it repelled electrons (like charges repel, see page 152): no current flowed. That is to say, this tube acted as a valve (see page 69), allowing electrons to flow in one direction only. It could therefore be used to convert an alternating current (a.c.), produced by an alternating voltage (see page 310), to a direct current (d.c., see Figure 32.3). This process is called rectification.

Fleming's valve, because it had two electrodes (an anode and a cathode) was called a diode (*di* = two). Whereas electrical devices control the

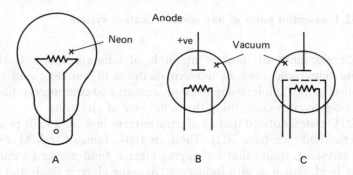

Figure 32.2 (A) an electric lamp (drawing); and diagrams representing (B) a Fleming diode, and (C) a triode valve

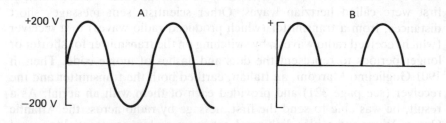

Figure 32.3 (A) alternating voltage supply produces an alternating current; and (B) half-wave rectified direct current

movement of electricity in wires (see Chapter 14), the Fleming diode controlled the movement of electrons through a vacuum (see Figure 32.2) and was the first electronic device. The study of electronic devices (also called electronic components) and their use in electrical circuits, is called electronics.

Another kind of vacuum tube is called a triode valve (*tri* = three) because it has an electrically charged grid between the heater filament and the anode. A small change in the electrical charge on the grid has a much greater effect on the rate of flow of electrons from the cathode to the anode. This kind of valve can therefore be used in a radio to amplify (make bigger) the fluctuations in a weak radio signal received via the aerial.

In the 1940s a group of scientists in the USA investigated pure substances (for example, crystals of germanium and silicon) that were very poor conductors of electricity. They discovered, by accident, that these crystals were much better conductors if they contained very small amounts of certain impurities (for example, germanium crystals with a trace of arsenic, and silicon crystals with a trace of phosphorus or boron). These new conducting materials, because they are still poor conductors when compared with metals, are called *semiconductors* (Latin: *semi* = half).

Electronic devices made from these solid semiconducting materials are called *solid–state devices*, to distinguish them from the vacuum tubes which for most purposes they have replaced. Advantages of the solid–state devices include the following: they are much smaller, require a lower voltage and use less current, are easier to mass produce, are less expensive, and are more durable than Fleming diodes and similar vacuum tubes. Also, they work at room temperature: there is no heater and so no warm-up time before they start to work. The symbols used in circuit diagrams for four kinds of solid–state devices are illustrated in Figure 32.4.

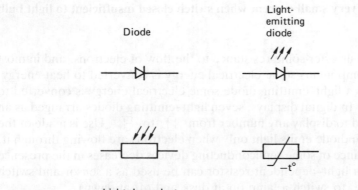

Figure 32.4 circuit symbols for four solid–state devices

A solid–state diode is made from two different materials (labelled p and n in Figure 32.5) with different conducting properties. Like a Fleming diode it allows a current to pass in one direction only and so can be used as a rectifier (see Figure 32.5). The arrow head in the symbol for a solid–state diode indicates the direction of conventional current flow, although you know electrons actually move in the opposite direction (see page 302).

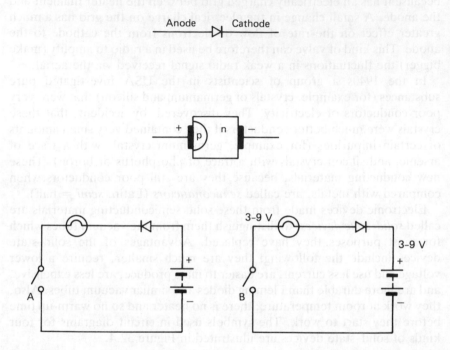

Figure 32.5 circuit symbol for solid–state diode, simplified cross section, and use in circuits: (A) bulb lights when switch closed; and (B) very small current when switch closed insufficient to light bulb

All materials offer some resistance to the flow of electrons, and in most wires and components some electrical energy is converted to heat energy. However, in a light–emitting diode some electrical energy is converted to light energy. In digital displays, seven light–emitting diodes arranged as an 8 are used to display any number from 0 to 9 . Use is made of the fact that each diode emits light only when electrons are flowing through it.

The resistance of some semiconducting devices decreases in the presence of light. So a light–dependent resistor can be used as a sensor and switch (for example, to switch a lamp on at dusk and off at dawn).

Whereas the resistance of other conductors of electricity increases with temperature, the resistance of semiconducting materials decreases with temperature. Semiconductors can be used as temperature–dependent resistors (called thermistors). The $-t°$ in the symbol for a thermistor (see Figure 32.4), indicates that the electrical resistance decreases as temperature increases. A thermistor can be used as part of a circuit to measure temperature (instead of a glass–liquid thermometer); and can be used as a sensor and switch (instead of a bimetallic strip) to switch on a heater — for example, as a room cools below the required temperature.

A transistor has three layers of semiconducting materials (B, C and E: see Figure 32.6). A small increase in the base current greatly increases the current flowing in the collector circuit. A transistor can therefore be used as an amplifier (like a triode valve, see page 364). It can also be used as a switch (off when there is no current or a very small current in the collector circuit, and on when a much larger current flows in the collector circuit).

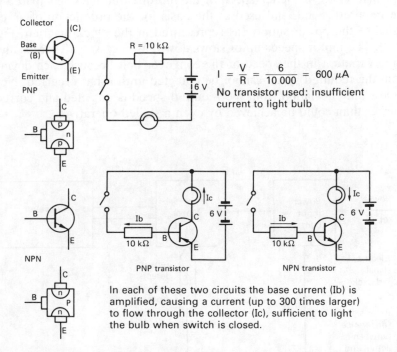

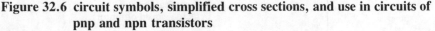

Figure 32.6 circuit symbols, simplified cross sections, and use in circuits of pnp and npn transistors

Some effects of feedback in control systems

If a sensor in a control circuit is placed too near the equipment it is controlling, feedback from the equipment may have an undesirable effect. For example, if a light–dependent resistor is used to switch on a lamp, the light from the lamp shining on this resistor may decrease its resistance — causing the lamp to go out. The light would therefore switch on and off repeatedly. This can be avoided by placing the sensor where the light from the lamp will not shine on it.

Feedback is desirable in many situations. For example, if a temperature–dependent resistor or a bimetallic strip is used as a sensor, and to switch on a heater when the room temperature falls below the desired level — then the same sensor will also switch off the heater as soon as the room temperature reaches the required level.

Feedback arrangements can be used to control almost anything that can be measured electronically by a suitable *transducer* (a device that converts one form of energy to another: for example, mechanical energy to electrical energy, see Figure 32.7).

In the automatic control of a conveyor belt in a workshop or factory, information on the actual speed of the motor is fed back, compared with the required speed, and used as the basis for the production of an error signal to the power supply (as represented in the block diagram, Figure 32.7). The motor speeds up or slows down, as necessary. There is always some variation in the speed of the conveyor belt because each deviation from the required speed cannot be detected until it has occurred. Nevertheless, any deviation from the required speed is detected and corrected quicker than could be achieved by even a skilled operative.

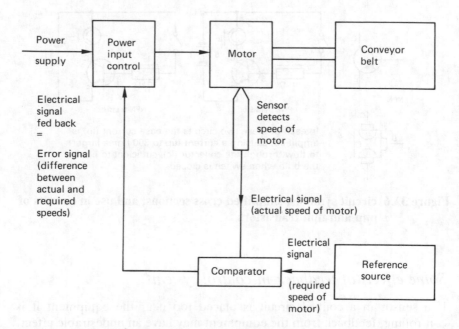

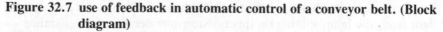

Figure 32.7 use of feedback in automatic control of a conveyor belt. (Block diagram)

32.2 Digital electronics

In digital electronics the only digits used are one and zero, because these can be represented by switching a circuit on (for one) and off (for zero) or by a high and a low voltage. The method of counting in which only these two digits are used is called the binary system (Latin, *bi* = two). Contrast this with the method of counting in which ten digits are used (0 to 9) which is called the denary or decimal system (*deca* = ten). If you are not familiar with binary numbers, try counting up to 15 on four fingers of one hand (with the first finger raised to represent one, the second only to represent two, then the first and second together to represent three, and so on: see Table 32.1). What number could you count up to, in the binary system, if you used all five digits of one hand?

Table 32.1   *Binary and equivalent denary (decimal) numbers 0 to 15*

Binary number	Explanation	Decimal number
0 0 0 0	0+0+0+0	0
0 0 0 1	0+0+0+1	1
0 0 1 0	0+0+2+0	2
0 0 1 1	0+0+2+1	3
0 1 0 0	0+4+0+0	4
0 1 0 1	0+4+0+1	5
0 1 1 0	0+4+2+0	6
0 1 1 1	0+4+2+1	7
1 0 0 0	8+0+0+0	8
1 0 0 1	8+0+0+1	9
1 0 1 0	8+0+2+0	10
1 0 1 1	8+0+2+1	11
1 1 0 0	8+4+0+0	12
1 1 0 1	8+4+0+1	13
1 1 1 0	8+4+2+0	14
1 1 1 1	8+4+2+1	15

Note that a denary (decimal) number can be converted to a binary number by repeatedly dividing the denary number by 2 and recording a 1 or a 0 at each stage to indicate the remainder. For example:

31 15 7 3 1 0
    1   1 1 1 1

and

75	37	18	9	4	2	1	0
	1	1	0	1	0	0	1

All telephone systems used to be *analogue systems*. An analogy is when one thing can be compared with something else that is actually quite different. As you speak into an analogue telephone system — many are still in use — an audio signal (the continuous and varying sounds of each word spoken) is converted to an electrial signal (which is continuous and varying in voltage, mimicking or analogous to the varying soundwaves). This *analogue signal*, carried a long distance in a telephone wire, is subject to distortion and interference. The telephone receiver converts this electrical signal into an audio signal that resembles your voice more or less closely (depending on the amount of distortion of the electrical signal) and is accompanied by background noise (the unwanted signals resulting from interference).

In telephones using a *digital system* your voice is sampled many times each second. Each sample of sound is then transmitted as a different electrical signal that represents a binary number. At the receiver this *digital signal* is converted to a sound. All the sounds produced are clear. They resemble the sounds of your voice very closely. There is no discernable distortion and no background noise: nothing is added to the signal during transmission. You will understand that a digital telephone system can transmit 16 different sounds encoded as 16 different patterns, each comprising four binary digits (see left-hand column of Table 32.1). The receiver's exchange decodes the 16 patterns into the original 16 signals (sounds).

The first computers were analogue instruments, made to enable scientists and engineers to perform arithmetic calculations that were too time-consuming or too complex to attempt by other methods. Most modern computers are digital instruments, and are also used to perform calculations.

However, digital computers can do much more than this: they can sort, file and analyse information rapidly. The binary system is used to represent not only numbers but also words, thoughts and actions. Before it can be processed or stored, all information (the input to the computer) must be converted to a binary code — with numbers represented by electrical signals (one pulse of electricity = 1, no pulse = 0; or high voltage = 1, low voltage = 0).

In a digital computer, one circuit off or on provides two binary digits (0 and 1). Two circuits provide 2×2 arrangements of binary digits (00, 01, 10, and 11). Three circuits provide $2 \times 2 \times 2$ arrangements (see Table 32.1). So, n circuits would provide 2^n arrangements.

In computing, *each digit is called a bit*. By international agreement eight bits are used to provide 256 different arrangements of digits (bit codes) to represent the letters of the alphabet, digits, punctuation marks and other characters. That is to say, eight bits are used to provide the particular arrangement of digits that represents each letter. For a two letter word 16 bits are needed; for a three letter word 24 bits; and so on. *Eight bits are commonly referred to as one byte*, and the storage capacity of a computer is usually expressed in bytes.

The electronic devices used to control the flow of information through a computer, and through many other electronic instruments, are called logic gates. They are called gates because they can only be either open or closed; that is to say they are two–state electronic devices. Each gate is an integrated circuit constructed from resistors, diodes and transistors.

Most logic gates have two inputs (A and B) and one output (see Table 32.2). The output from a gate is either a signal (a high voltage, called logic one, when the gate is open) or no signal (a low voltage, called logic zero, when the gate is closed).

One type of gate, called an AND gate, is open (output voltage high = logic 1) only when both inputs (A AND B) are connected to the positive terminal of the battery. Then both inputs are high (at logic 1, see Table 32.2). The circuit symbols for five types of logic gates, with the possible inputs and outputs, are included in Table 32.2.

Another type, the NOT gate, has only one input. Its output signal is always the opposite of the input signal (see Table 32.3). A NAND gate can be made from an AND gate followed by a NOT gate, and a NOR gate from an OR gate followed by a NOT gate. In the circuit symbols, the circle (o) always indicates NOT (as in Table 32.2).

In a computer different arrangements of logic gates are used to calculate (by addition and subtraction), and to store and compare data. For a digital system to be called a computer it must be able to receive information (data) in a binary code, store data, calculate, make decisions, and produce an output.

32.3 The impact of information and control technology on everyday life

The development of very small solid–state electronic devices has made possible the construction of minicomputers, and of the microcomputers and pocket calculators used in small businesses, schools and many homes. As a result, the calculations that can be performed with a large computer constructed in the 1950s (occupying about seven cubic metres) can now be performed with a programmable pocket calculator made in the 1970s. Similarly, the time indicated by an analogue display on a clockwork watch could by the 1970s be indicated by a digital display on an electronic watch.

The silicon chips used in computers and calculators are thin pieces of silicon a few millimetres square. Each of these can contain many solid–state devices (resistors, diodes and transistors). A single chip is called

Table 32.2   *Truth table and circuit symbols for five types of logic gate, each with two inputs (A and B) and one output* (o/p)

Possible inputs		Outputs from each type of gate				
A	B	And gate	Nand gate	Or gate	Nor gate	Ex-or gate
0	0	0	1	0	1	0
0	1	0	1	1	0	1
1	0	0	1	1	0	1
1	1	1	0	1	0	0

Circuit symbols:

	And	Nand	Or	Nor	Ex-or
American (old)					
IEC/British (new)					

Note. And gate: output 1 only when both inputs = 1
Nand gate: opposite of And gate
Or gate: output 1 when one or both inputs = 1
Nor gate: opposite of Or gate
Ex-or gate: output 1 only when inputs are different

Table 32.3   *Truth table and circuit symbols for a NOT gate*

Input	Output	Circuit symbols	Not
A	Not gate		
0	1	American standards (old)	
1	0	International Electrotechnical Commission/British (new)	

an integrated circuit, and is enclosed in a larger protective case through which wires pass for electrical connections to other components.

The increasing use of computers is one aspect of the electronics revolution that is affecting your life. Computers are used, for example, in offices, factories, hospitals, shops and homes. In a shop the cost of things you buy may be added up using a computer; and at the same time the computer may amend stock records and inform the ordering department when more goods of any kind are required. In schools computers may be used as aids to instruction and learning; and many people have computers they use for education and pleasure as well as for record keeping in their homes. In addition, many household items (some washing machines, vacuum cleaners, refrigerators and cameras, for example) contain a small computer (perhaps a single–chip microprocessor, with a number of integrated circuits on one silicon chip).

Computers have made possible the development of word processors, digital telephones, electronic mail (including fax = facsimile), and via telephone links the use of other people's computers and databases (information stored in computers). These changes in our ability to communicate have enabled some people to work from home, instead of travelling to work, and have also enabled many handicapped people to do work which previously they could not have attempted.

In offices and in industry, computers are being used to do many of the tasks that previously could be done only by people. The use of equipment that does things automatically, requiring people only for maintenance, servicing and supervision, is called automation. Some machines that do work previously done by people are called robots.

Computerisation in an office can result in tasks being undertaken that were previously too time–consuming or too complex. As a result more people may be employed but costs may be reduced in other ways (for example, by better stock control).

Robots in industry, and other aspects of automation, should reduce the number of people required. Many repetitive tasks can be performed better if machines are controlled by computers instead of by people. As a result there may be no manual labourers and few craftsmen in some factories where previously many were employed.

Investigation 32.1 *Information technology*

Experience in the use of microelectronic equipment for information transfer, information processing and information storage can be obtained only by using the equipment. Because of the importance of information technology in so many aspects of life today, as well as in science, you are advised to take any opportunity to gain such hands–on experience. Use the instructions provided with the equipment and if possible join a class so that you can ask questions and receive expert tuition.

Table 32.4   *The electromagnetic spectrum*

Names of waves	Wave length (m)		Some uses
gamma rays	10^{-14}	*shortest wavelength highest frequency*	Sterilising instruments
X-rays	10^{-18}		X-ray photography
ultraviolet rays	10^{-8}		
violet indigo blue green yellow orange red		visible light	lighting photography communication
infrared waves	10^{-5}		electric fires
microwaves	10^{-2}		cooking communications
radar waves	3×10^{-2}		air traffic control UHF television
radio waves		*longest wavelength*	VHF radio, medium-wave radio,
	2000	*lowest frequency*	long-wave radio

Note. Each type of radiation has a range of wave lengths. Together, the radiations form a continuous spectrum. They all consist of oscillating electric and magnetic fields, are transverse waves, and travel through a vacuum at the speed of light (3×10^8 metres per second).

One result of the development of new methods of communication, for example telegraphy in 1837, the telephone in 1875, radio in 1901, television in the 1920s and electronic mail (including facsimile = fax) in the 1980s, is that people in each country soon know of events in other parts of the world.

33.1 Always more people

The first people hunted wild animals and gathered plant food. Their numbers were probably limited by the amount of food available in the most difficult season of the year. When people started to domesticate wild animals (about 15 000 years ago), and to cultivate the land (about 10 000 years ago), they increased the amount of food available throughout the year. These changes made possible the survival of more people. Since then each agricultural improvement has made further population growth possible.

It took more than two million years for the population of the world to reach 1000 million (in about 1800) but only another 100 years for the population to reach 2000 million (in about 1900). Another 1000 million were added in the next 60 years (total 3000 million in about 1960); another 1000 million in the next 15 years (total 4000 million in about 1975); and another 1000 million in the next 12 years (total 5000 million in about 1987). Each million has been added in a shorter time.

Population growth at the rate at which it has been taking place in the last two hundred years has had two basic causes:

1 The results of scientific research applied in medicine.
As a result of preventive medicine (community or public health measures, contributing to disease control), many people who would previously have died in infancy or childhood now survive and have children. Since the first vaccination against smallpox by Jenner in 1798 and Pasteur's immunisation of sheep against anthrax in 1881, the techniques of vaccination and immunisation have been developed and used in the prevention of many diseases caused by viruses and bacteria which used to kill many people. Since 1897, when Ross discovered malarial parasites in the *Anopheles* mosquito, control measures against mosquitoes and the development of anti-malarial drugs, have resulted in the eradication of malaria in some places and a reduction in the number of deaths caused by malaria in other

places. Since 1928, when Fleming concluded that the fungus *Penicillium* secreted a chemical that was toxic to bacteria, penicillin and many other antibiotics have been produced. As a result it is now possible to treat and cure many diseases caused by bacteria.

2 The results of scientific research applied in agriculture.
Yields from crop plants have been increased following the development and use of fertilisers to provide the nutrients required by the plants. Similarly, yields from farm animals have been increased as a result of our better understanding of their food needs. Discoveries in genetics, applied to plant and animal breeding, have also resulted in increased food production. And studies of the pests and diseases affecting crop plants and farm animals have made possible the control of many pests and the prevention of many diseases, and so contributed to increased food production.

Population growth started in pre-history and continued despite the ravages of disease and deadly quarrels. Recent improvements in the prevention and cure of many diseases of people, as a result of the application of the results of scientific research, have contributed to rapid population growth. However, without the application of the results of scientific research to agricultural improvement in the past 100 years it would not have been possible to support the extra people. As long as more food is available, more people are likely to survive, but they will not necessarily be better fed. Unfortunately, although increasing food production allows more people to survive it does not prevent food shortages. Increasing food production and population growth also have adverse effects on the environment.

33.2 Environmental impact and cost–benefit analyses

When you are thinking of buying something, you consider how much it costs and whether it is worth it. The same kind of cost–benefit analysis can be applied to any activity; but — unfortunately — although some benefits may be obvious many of the costs may not be. Furthermore, short–term benefits may be followed by long–term costs, and the benefits to an individual may be followed by costs to society.

For example, people find aerosols useful for many purposes (for example, for spraying perfumes, paints and pesticides). They buy them because they think they are worth the purchase price. That is to say, they think the benefit is greater than the cost. Unfortunately, the aerosols introduced in the 1960s, and still used in the 1990s, contained CFCs that contributed to the depletion of the ozone layer (see page 275) and to the atmospheric pollution that enhances the greenhouse effect (see page 275). These environmental costs were not considered by most people when they bought aerosols.

People do many things that damage the environment. Any benefits resulting from these activities should be considered in relation to the environmental and social costs. It is best if an analysis of environmental and social impact is undertaken, before any activity is started, in an attempt to determine the likely benefits and possible environmental and social costs. But it is easier to appreciate the benefits and costs if, in the light of experience, we consider activities that people have been engaged in for some time.

Nuclear fission is potentially a source of great power. So when scientists succeeded in controlling nuclear fission and the first nuclear power stations were built in the 1950s, it is hardly surprising that people thought energy from nuclear reactors would provide a pollution–free and unlimited energy source, as an alternative to fossil fuels. However, the development of nuclear power stations has been slowed down in the 1990s in some countries following contamination of the environment with harmful radioactive materials after accidents at existing nuclear power stations. There has also been public concern about the disposal of radioactive waste materials, with a long half-life, from nuclear power stations.

The cells of living organisms can be damaged by radiation from radioactive materials. When a nuclear fission bomb was dropped on Nagasaki, Japan, in 1945, about one fifth of the 130 000 people who died were killed by the effects of radiation. In nuclear power stations, radioactive materials are kept inside the reactor vessel. Any radiation passing through the steel is trapped in a layer of concrete about 3 m thick. (The diagram of a reactor (Figure 29.4) is simplified to represent the method of operation, not details of the actual construction.) The radioactive material is used as completely as possible; but a small amount has to be disposed of safely by long–term storage (see half-life, page 335).

Even natural background radiation can induce cancers and cause genetic damage affecting future generations, and the harmful effects of radiation are increased by any environmental pollution with radioactive materials resulting from nuclear explosions and from emissions from nuclear power stations following accidents.

In a cost–benefit analysis the harmful effects of radiation should be balanced against the use made of radioactive materials, not only in generating electricity but also in medicine, industry and research. For example, in medicine gamma rays are used to kill cancer cells and to sterilise instruments; and radioactive isotopes are used as tracers — to help doctors follow the movement of labelled materials through the body of a patient. In industry gamma rays can be used to measure the thickness of a material as part of a control system, and radioactive isotopes can be used as tracers — to help investigators follow the movement of labelled materials through underground pipes.

Environmental impact of agriculture and medicine

Population growth, resulting from increased food production, preventive medicine, and advances in the treatment of disease, has resulted in attempts to produce more and more food (see Figure 33.1).

Since the start of cultivation, one way in which people have increased food production has been by increasing the area of land under cultivation. Obviously, this has resulted in a corresponding decrease in the area covered by natural vegetation — and so in a progressive decrease in animals' habitats. By the spread of agriculture, more than in any other way, people have destroyed natural ecosystems and altered Earth's surface.

The seemingly obvious benefits of agriculture and medicine, in making possible the survival of more people, may be considered in relation to these and other environmental and social costs.

The destruction of natural vegetation, although largely due to the spread of agriculture, has also resulted from an increasing demand for timber for building and furniture making, for wood pulp used in making paper, and for wood as a fuel. When forests are cut down faster than trees are replaced, the result is deforestation.

The environmental cost of deforestation on hillsides, unless a cover of vegetation can be maintained or other methods are used to retain the soil, is soil erosion. Other environmental costs resulting from soil erosion include the silting of lakes and rivers (and manmade reservoirs), and the extension of coastal mud flats. Furthermore, in the absence of soil, water is not retained on the hills and there is an increased risk of floods in the lowlands. Without soil farming becomes impossible in the hills; and in the lowlands the floods ruin crops and make people homeless.

Deforestation, inappropriate cultivation techniques, and overgrazing, on desert margins, are also resulting in soil erosion and desert expansion. This is occurring particularly in the tropics, in regions where large numbers of people are making greater demands upon the productivity of soils — with the result that soil fertility is decreasing. With larger populations, less food is being produced. Deforestation, overcultivation and overgrazing, soil erosion, desert expansion, and reduced productivity of the land, are some of the environmental costs of population growth.

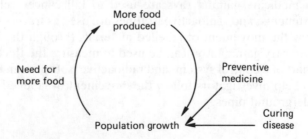

Figure 33.1 Food for people

Environmental pollution by agriculture and other industries

Agriculture is one cause of pollution. The farmer has no control over the rate at which chemicals used on crops or on the soil are washed into streams, rivers and lakes. This kind of pollution may at first favour the growth of algae in the water. But when the algae die and are decomposed by saprobionts, the oxygen in the water may be used completely. This kills all other organisms. The changes leading to this lifeless condition are called eutrophication. Water contaminated with chemicals also percolates through the soil and pollutes the water table. As a result underground water sources may become unfit for use as drinking water.

Pesticides are used to kill organisms that are harmful to our interests. Unfortunately some beneficial organisms may be killed at the same time. For example, insecticides may kill natural predators and parasites of the pest insects as well as the pests themselves. They also kill insects in the soil that play a beneficial role in the decomposition of humus; and they kill honey-bees and other insects that are essential for the pollination of some crop plants.

Another problem with the use of pesticides is that qualities which are desirable in some ways may be undesirable in others. For example, a pesticide that is not broken down quickly is a persistent toxic chemical — it will continue to be effective as a pesticide for some time after it has been applied (see page 353). However, persistent chemicals can accumulate in the bodies of other organisms, especially in animals that come later in food chains, if at each stage the organisms consume other organisms that contain the persistent chemical. At these higher concentrations a pesticide may be harmful to other organisms as well as to the pests against which it was used.

Apart from agriculture, many other industries produce harmful chemicals which may pollute the environment. The problem of accumulation, with the concentrations of toxic chemicals increasing at each stage in a food chain, occurs with some of the chemicals present in effluents from factories. For example, heavy metals such as mercury persist in the environment and may accumulate in food chains. So the dilution of an effluent does not provide a solution to all pollution problems. Foods containing a high concentration of such toxic chemicals may be unfit to eat.

Coal–fired power stations (see page 326), blast furnaces (see page 289) and factories burning coal, produce smoke and colourless fumes which pollute the air. These pollutants irritate the surface of the lungs and are harmful to health (see page 213). They are also harmful to other organisms; as when sulphur dioxide in the atmosphere combines with water and results in the acid rain that is suspected of causing the death of trees in some forests and of fishes in some lakes.

It is feared that pollution of the upper atmosphere is damaging the ozone layer (see page 275) and resulting in the accumulation of greenhouse effect gases (see page 275) which may change not only the climate but also the area of Earth covered by ice, cause an increase in sea level, permanent flooding of many lowland and coastal areas and changes in Earth's climate.

One way in which the polluting effects of agriculture could be reduced is by adopting agricultural methods similar to the traditional methods used until recently in each part of the world.

In a factory, the smoke or fumes from a chimney, and the liquid effluent into a drain or river, is localised and can be treated. The cost of pollution control to the manufacturer results in an increased cost of the product to the consumer but this may be offset by the reduced environmental and social costs of air and river pollution.

Environmental impact of urbanisation

In all parts of the world the decline in peasant agriculture, the increase in industrial development near towns, and the prospect of higher earnings in towns, are resulting in the depopulation of rural areas, the growth of towns into cities, and the merging of towns and cities in vast urban areas called conurbations.

The growth of the urban area reduces the land available for farming, especially because urban development takes place in lowland areas, covering land that is most suitable for agriculture.

Urbanisation is an obvious example of the environmental impact of population growth and the concentration of more people into a smaller area. The costs of urbanisation include not only the great cost of maintaining urban societies, and the demands put on nearby and distant rural areas, but also the great cost of maintaining the health of people in crowded and perhaps polluted conditions.

32.3 Too many people?

The word conservation means different things to different people. To a biologist, nature conservation means managing natural communities so a variety of organisms can live in their natural surroundings with little interference from people. Considering the world as a whole, as one ecosystem, conservation is the wise use of resources in an attempt to minimise the harmful effects of people upon other organisms and upon the land, rivers, lakes and seas.

In clearing the land for agriculture, people destroy natural communities. As a result of bad farming people have already made some fertile areas of the world unsuitable for agriculture. Our priorities in agriculture should be first to conserve the soil and second to maintain or improve soil fertility. Yet in many parts of the world desert encroachment makes land un-productive. We are desert makers.

People were hunting and fishing before the start of cultivation; and some people have continued with these activities. But with the growth of populations and the development of more effective weapons, large mammals have been eliminated in many places. In other places wild animals are conserved and cropped. They are a source of food and other useful materials, and also a tourist attraction. They are worth conserving even if only as a source of revenue.

Similarly, the seas were once thought to contain so many fishes that their numbers could not be much reduced by fishing. But with improvements in fishing techniques and with fishing fleets of many nations competing to remove more and more fish from the same seas, overfishing has taken place in many fisheries. Overfishing is the removal of so many fish in one year that there are fewer in the following year.

The growth of populations also has a direct adverse effect on the land. Urbanisation makes land unproductive. People in towns and cities make demands on the surrounding countryside; building roads, canals, railways, ports and airports. They make demands on both nearby and distant lands as sources of food, as sources of the raw materials used in industries and to further the process of urbanisation, and as places where they can relax and enjoy themselves. These great accumulations of people destroy the species diversity of the area over which they sprawl. And they make the conservation of wild life more difficult in many other parts of Earth's surface.

People compete with other organisms for food and for a place in which to live. We are the most destructive creatures ever to have lived on this planet. Population growth makes the conservation of natural resources more and more difficult.

How then can people conserve other organisms, the soil, and other limited resources, so that the world remains fit for people to live in? Legislation may help. Laws concerning the output of wastes from factories help to reduce such things as smoke production and the release of toxic chemicals. By law certain areas may be set aside as nature reserves or national parks, in an attempt to reduce the impact of people. Attempts are also made, by international agreement, to conserve stocks of fishes and whales — for example, by allowing fishing only in certain seasons or by banning the catching of certain species.

Population growth not only makes the conservation of natural resources increasingly difficult, but also affects human life directly. Each year we need more food and drinking water, and more houses, schools and hospitals, than the year before. And each year more waste is produced.

Education is important, to let people know that population growth makes conservation more difficult and aggravates many economic and social problems. Indeed, in some countries there are advertising campaigns to encourage people to have smaller families (see Figure 33.2).

In thinking of the future, population growth should not be accepted as inevitable. We should not try to fit as many people as possible into the

Figure 33.2 A family planning poster in India: a mother, father and two children — and the words 'A small family is a happy family'. In the foreground the parents of a large family are speaking to a family planning worker. (Photograph courtesy of International Planned Parenthood Federation.)

world, if this would mean increasing the number of homeless, illiterate and unemployable people. It would be better to have fewer people if this would mean healthier people leading a better life.

People should respect other forms of life; and so should not plan only for people. Anyone who is not convinced of the need to respect and conserve other organisms and the places in which they live, should remember that all the world is one ecosystem. All organisms are interdependent. People do not, and could not, exist by themselves.

◯ Questions

The following questions and advice are to help you if you are preparing for an examination. Examples of different types of question are included; but you are advised to obtain copies of recent papers set by the Examination Board whose examinations you will be taking. They are your only reliable guide to the kinds of question, and to the number of questions, likely to be in the papers you take.

The most common fault in examinations is failing to answer precisely the question set. You cannot answer a question properly if you do not read it carefully to make sure you understand exactly what the examiners want to know. Another reason for reading each question carefully is that many questions provide information that will help you prepare your answer.

You must also read the instructions at the beginning of each question paper, before the first question. You will probably be instructed to answer all questions on the answer sheet provided. You may be advised not to spend too long on any question, but to work through the paper answering first the questions you feel confident you can answer correctly. You can then work through the paper again, if you have time, tackling the questions you have not already answered. This is to make sure you do answer all the questions you are most confident about.

You should do your best to answer all questions. In the instructions you may be advised that no marks will be deducted for wrong answers. This means, if you are not sure of the correct answer, you should guess. You will score marks if your guess is right.

1 Some questions test your ability to observe carefully, measure precisely, record accurately, and perform simple calculations. For example, what is the volume of the irregularly shaped object represented in Figure 4.2?

2 Some questions are set to find out if you can explain some of the things you have observed. For example, in Figure 4.2 the observer can see the water surface as two parallel lines. Why does water cling to the jar, forming a meniscus?

3 In some questions you may be tested to see if you know how to use tables and graphs. For example, from Figure 5.8, a graph, estimate the length of the tail of the mouse after: (a) 20 days, (b) 30 days, and (c) 40 days.

Completion questions

4 Given that the atomic number of chlorine is 17 and its mass number is 35, complete Table A1.1.

Table A1.1

	Chlorine atom Cl	*Chloride ion* Cl⁻
Number of protons	17	
Number of electrons	17	
Number of neutrons		18

5 Complete the circuit diagrams (Figure A1.1 A, B, C and D) so that in A the light is on when the switch is closed; in B the light is on when either switch is closed; in C the light is on only when both switches are closed; and in D both lights are on when the switch is closed.

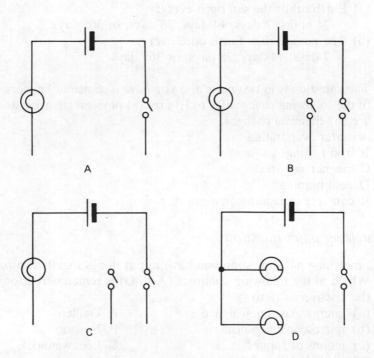

Figure A1.1 **incomplete circuit diagrams**

6 In the diagram representing magnets and magnetic fields (Figure A1.2), if W is a north pole, what are the other three poles:

X =   ; Y =   ; and Z =

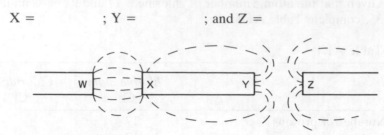

Figure A1.2 magnetic fields

Multiple choice questions

In a multiple choice question you have to choose the correct answer from the selection of answers provided.

7 Which ending is needed to make a correct statement in each of the following.
   (a) Earth orbits the sun once every:
        24 hours, 7 days, 14 days, 28 days, or 365 days.
   (b) The moon orbits Earth once every:
        7 days, 14 days, 28 days, or 365 days.

8 This question is in two parts and you have a choice of answers. Which of the following changes (A to E) are: (a) physical changes, and which are (b) chemical changes.
   A water evaporating
   B iron rusting
   C magnetising steel
   D coal burning
   E concrete expanding on a hot day

Matching pairs questions

In a matching pairs question you have to put things together in pairs.

9 Which of the following scientists (A to G) is remembered for each of the discoveries (a to f)?
   (a) energy/mass equivalence   A Galileo
   (b) microscopic organisms   B Newton
   (c) moons of Jupiter   C Leeuwenhoek
   (d) white light is a mixture   D Mendel
       of coloured lights   E Mendeleev
   (e) law of segregation of alleles   F Einstein
   (f) periodic classification of elements   G Pasteur

10 Which of the materials (A to F)
 (a) is a good conductor of electricity   A oxygen
 (b) is a compound   B sulphur
 (c) reacts with water, releasing   C copper
     hydrogen gas   D sodium
 (d) burns in air giving an acidic oxide   E air
 (e) is a mixture   F water

Short answer questions

In a science examination you will probably be provided with either an answer sheet or a question paper in which there is space after each question for your answer. But, in this book, do not write your answers in the spaces provided. Use a separate notebook, but note that the amount of space left for each answer is an indication, as in an examination, of the length of answer the examiners require. Think, then answer directly — as clearly and simply as you can. You will not score extra marks for giving more information than is necessary. Unnecessary words just make your answers harder to mark.

11 Table A1.2 includes some information about a 2 cm cube. Copy this table in your notebook, then complete the table by adding the surface area, volume, and surface to volume ratio, of a 4 cm cube and of a 1 cm cube.

Table A1.2

	Surface area	Volume	Ratio Surface: volume
4 cm cube			
2 cm cube	24 cm^2	8 cm^3	3:1
1 cm cube			

(a) What do you conclude from your completed table about the relationship between surface area and volume?

(b) When a cell divides into two (see page 94) its volume is halved. Is its surface area halved?

(c) When you chew food does its total volume: (i) increase, (ii) remain the same, or (iii) decrease?

(d) Does the food's total surface area: (i) increase, (ii) remain the same, or (iii) decrease?

(e) After you have been chewing bread for some time it begins to taste sweet. Write a word equation to summarise the chemical reaction that is taking place.

(f) Explain how chewing bread increases the rate of this chemical reaction.

(g) Each enzyme works best at a particular pH, called the optimum pH for that enzyme. The digestion of starch occurs rapidly while food is being chewed but stops soon after the food reaches the stomach. Suggest a reason for this.

(h) Define the term *catalyst*.

Comprehension questions

In some questions you are given information on a topic you might not have studied previously. For example, in questions 12, 13 and 15 you are given information that is not included elsewhere in this book. You can answer these questions if you consider carefully the information provided in each question — and your knowledge of science gained from reading this book.

12 Figure A1.3A represents a piece of dialysis tubing, attached to a glass tube, and containing a sugar solution. This is immersed in a beaker of water. Dialysis tubing has very small pores through which water molecules can pass but not the larger sugar molecules. It is called a differentially permeable membrane. Water molecules diffuse through this membrane into the sugar solution — where they are in lower concentration. This movement of water is called osmosis. As a result of osmosis, the sugar solution is diluted and rises in the glass tube.

Figure A1.3B represents two cells in the root of a plant, a microscopic vessel or tube that passes through the root and stem of the plant and into a leaf, and two cells of the leaf. Many surface cells of the root take up water by osmosis, and there is a continuous column of water through the root and stem and into the leaves — from which, whenever the stomata are open, water is lost by evaporation. It is the evaporation of water from the leaves that pulls more water up, through the water-filled vessels, to the leaves.

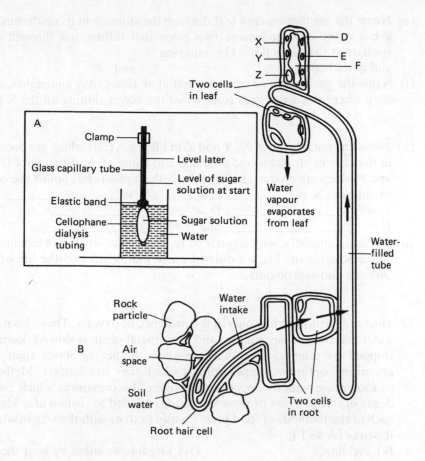

Figure A1.3 (A) demonstration of osmosis and osmotic pressure;
(B) movement of water through a plant, represented by
arrows

(a) What is diffusion?

(b) What is the force of attraction between molecules called, which results in the molecules holding together as water is pulled up the vessels to the leaves?

(c) How do you think the evaporation of water affects the temperature of a leaf on a hot sunny day?

(d) Explain why the evaporation of water has this effect.

(e) Name the gas that enters a leaf through the stomata in its epidermis on a hot sunny day; and name two gases that diffuse out through the stomata at the same time. Gas entering =
and gases leaving =   and

(f) Name the gas used in all cells of a leaf at all times (day and night), and state where this gas comes from when the sun is shining on the leaf.

(g) Name the parts labelled X, Y and Z in Figure A1.3B which are present in the cells of all plants and animals; and name each of the parts D, E and F which are characteristic of the cells of plants but not of the cells of animals: X = _____, Y = _____,
Z = _____; D = _____,
E = _____, and F = _____.

(h) Explain, concisely, why directly or indirectly the structures labelled E make possible the life not only of green plants but also the life of all animals and saprobionts.

13 Most organisms can live only in the presence of oxygen. They also need water and food to sustain life, and their metabolism is slowed down or stopped by temperatures that are either higher or lower than the organisms' optimum temperature at which they live longest. Methods of food preservation depend on depriving the organisms which cause decay of one or more of these conditions needed to sustain life. Match each of the methods of food preservation (a to e) with the way in which it works (A to F):

(a) Pickling.	(A) Organisms killed by heat then other organisms excluded.
(b) Making jam and similar preserves	(B) Metabolism of microbes slowed by low temperature, reducing or stopping their growth and reproduction.
(c) Drying in sun; and freeze-drying in industry.	
(d) Freezing or packing in ice.	
(e) Canning and bottling.	(C) High concentration of sugar; water withdrawn from microbes by osmosis.
	(D) High concentration of alcohol kills microbes.
	(E) Dehydration kills microbes, or prevents their growth and reproduction.
	(F) Low pH kills microbes.

14 Explain why, (a) in a shop or in your home, cooked meat should not be kept where it could come into contact with uncooked meat; (b) a utensil used to cut or serve cooked meat should never be used to cut uncooked meat; and (c) a person handling or serving cooked meats should not handle uncooked meat.

(a)

(b)

(c)

15 If pollen produced in the anther of one flower is carried to the stigma of another flower of the same species (for example, by an insect or by the wind), the flower is said to have been cross–pollinated. Insects may be attracted by the coloured petals, and may ingest nectar secreted by nectaries at the base of the petals. The pollen germinates, and a pollen tube (see Figure A1.4) grows through the tissues of the style. The pollen tube has two haploid nuclci, one of which fuses with the egg nucleus — which is also haploid. This will remind you of the fusion of a sperm with an egg in the reproduction of mammals. In plants, fertilisation is followed by the development of a seed; and the ovary wall develops into a fruit which contains all the seeds produced in one flower.

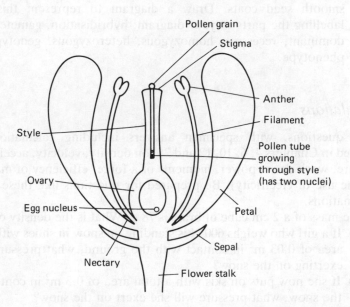

Figure A1.4 the parts of a flower (diagrammatic section)

(a) In the spaces provided, below, explain briefly the meaning of each of the following terms;

Species:

Ingest:

Secrete:

Self pollination:

(b) Complete the paragraph, below, using words selected from this list: haploid, diploid, zygote, mitotic, fertilisation, sexual, asexual, meiotic, chromosomes, and chromatids.

Reproduction involving the formation and fusion of gametes is called _____reproduction. The fusion of gametes is called_____ and the diploid cell formed is called a _____. This has twice as many _____ as a haploid cell. By_____ cell division it develops into a new multicellular individual of the same species, all the cells of which are _____.

(c) When Gregor Mendel crossed pea plants that were pure-breeding for smooth seed coats with other pea plants that were pure-breeding for wrinkled seed coats, all the seeds produced had smooth seed coats. Draw a diagram to represent this cross, labelling the parts of your diagram: hybridisation, gamete, allele, dominant, recessive, homozygous, heterozygous, genotype, and phenotype.

Calculations

Some questions, with specimen answers including calculations, are included in Chapters 4, 6, 10, 12 and 28 (on density, velocity, acceleration, pressure, work done, power, moments of a force, efficiency of machines, and the use of electricity). Be prepared for questions like these in your examinations.

16 The mass of a 2 cm cube of brass is 78 g. What is the density of brass?

17 (a) If a girl who weighs 600 N is standing on snow in shoes with a total area of 0.03 m^2 in contact with the ground, what pressure is she exerting on the snow?

(b) If she now puts on skis with a total area of 0.3 m^2 in contact with the snow, what pressure will she exert on the snow?

Assessed course work

The grades awarded at the end of most science courses are based partly on marks gained in externally assessed examinations, and partly on assessed course work. Course work includes practical investigations, like the investigations in this book, and to get good grades you must:

a Understand the purpose of each investigation.

b In an experiment, select appropriate materials and plan the investigation to test a hypothesis.

c Always follow instructions carefully and work safely.

d Observe carefully, measure precisely, record data accurately and analyse data correctly.

e Prepare clear, concise and accurate accounts of your investigations, using the usual headings (see page 8), including diagrams if these will help the reader.

f List conclusions and suggest possible further investigations.

Some questions set in theory examinations test your knowledge of apparatus used in science laboratories and of practical techniques. Such questions also test your ability to relate your theoretical studies to your practical investigations.

18 Explain, with the aid of a diagram, how you would determine the mass of salt in 1000 cm³ sea water, given enough sea water, a measuring cylinder, a tripod stand, a wire gauze, an evaporating dish, a bunsen burner, and a suitable balance.

19 Name a metal and an acid that could be used to prepare hydrogen gas, using apparatus similar to that illustrated in Figure A1.5.

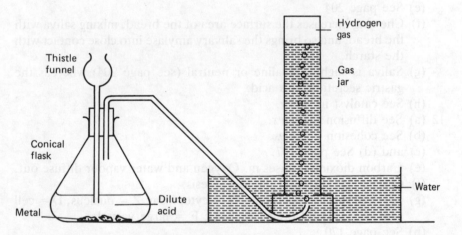

Figure A1.5 the production and collection of hydrogen: the gas displaces the water in the gas jar

Answers to questions

1 Calculate the value by subtracting the first reading from the second (4.5 − 3.0 = 1.5 cm³). Note that the answer is a number plus an SI unit of measurement.

2 See cohesion and adhesion in *Index*.

3 If you imagine a vertical line from 20 days to the line on the graph, and a horizontal line from this point to the vertical axis, you can estimate the tail length of the mouse aged 20 days.

4 The mass number of an atom is the number of protons plus the number of neutrons (see page 331). A chloride ion is negative because it has one more electron than a chlorine atom (see page 143).

5 (A) Draw horizontal line to complete circuit; (B) switches in parallel; (C) switches in series; and (D) bulbs in parallel. See *Index*.

6 Compare Figure A1.1 with Figure 27.1.

7 See Chapter 2.

8 (a) A, C and E; (b) B and D.

9 (a) F, (b) C, (c) A, (d) B, (e) D, (f) E.

10 (a) C, (b) F, (c) D, (d) B, (e) E.

11 (a) The surface area of a cube increases, in proportion to its volume, as its volume decreases.

   (b) No. Its surface area is greater in proportion to its volume.

   (c) (ii) Remains the same (unless it swells as a result of absorbing saliva).

   (d) (i) Increases.

   (e) See page 204.

   (f) Chewing increases the surface area of the bread, mixing saliva with the bread, and so brings the salivary amylase into close contact with the starch.

   (g) Saliva is slightly alkaline or neutral (see page 204) whereas the gastric secretions are acid.

   (h) See catalyst in *Index*.

12 (a) See diffusion in *Index*.

   (b) See cohesion in *Index*.

   (c) and (d) See page 107.

   (e) Carbon dioxide diffuses in. Oxygen and water vapour diffuse out.

   (f) Oxygen: produced in photosynthesis.

   (g) X = cell surface membrane; Y = cytoplasm; Z = nucleus; D = cell wall; E = plastid (a chloroplast); F = large vacuole.

   (h) See page 170.

13 (a) F, (b) C, (c) E, (d) B, (e) A.

14 (a) Microbes on uncooked meat may be transferred to cooked meat.

   (b) Utensils may be contaminated with microbes from uncooked meat.

   (c) Hands may be contaminated with microbes from uncooked meat.

15 (a) See *Index* for terms: species, ingest and secrete. The term cross–pollination is defined in the question, so you can work out that self–pollination must be the transfer of pollen from an anther of a flower to the stigma of the same flower.

   (b) The missing words, in order, are: sexual, fertilisation, zygote, chromosomes, mitotic, and diploid.

16 Density = 9.75 grammes per cubic centimetre.
17 (a) 20 000 Pa = 20 kPa.
   (b) 2000 Pa = 2 kPa.
18 **Note.** You do not need to use 1000 cm^3 sea water. You can determine the mass of salt in a sample (for example 50 cm^3 and then multiply your answer by 20 to calculate the mass of salt in 1000 cm^3).
19 See page 147.

◯ Further reading

Introductions to science subjects

Barrass, R. *Human Biology Made Simple*, Heinemann: Polybooks, Sunderland.
Barrass, R. *Modern Biology Made Simple*, Heinemann: Polybooks, Sunderland.
Critchlow, P. *Mastering Chemistry*, Macmillan, Basingstoke.
Keighley, H. J. P., McKim, F. R., Clark, A. & Harrison, M. J. *Mastering Physics*, Macmillan, Basingstoke.

Books on scientific investigation

Asimov, I. *Asimov's New Guide to Science*, Basic Books Inc., New York, and Viking/Penguin, Harmondsworth.
Beveridge, W. I. B. *The Art of Scientific Investigation*, Heinemann, London.
Barrass, R. *The Locust: a laboratory guide* 3rd edn, Heinemann: Polybooks, Sunderland. Includes an introduction to the locust problem, advice on rearing locusts, and suggestions for many investigations with insects.

Books on study, revision and examination techniques

Barrass, R. *Study: A guide to effective study, revision and examination techniques*, Chapman & Hall, London.
Clough, E. *Teach Yourself Study and Examination Techniques*, Edward Arnold, Sevenoaks.

The book by Clough is recommended to anyone preparing for school examinations taken at 16 plus, such as GCSE and the Scottish Certificate of Education Standard Grade, and similar examinations comprising short answer questions taken by older students.

The book by Barrass is aimed at students preparing for school examinations taken at 18 plus, or for more advanced examinations in further and higher education.

Reference books for your bookshelf

For anyone who takes study seriously, a good dictionary is an essential reference book. Buy one that gives the spelling, meanings, pronunciation, origin, and present status in the language of each word. It will be a life-long source of interest and pleasure.

Other books that will help you to improve your use of words, and so help you to organise and communicate your thoughts, are:

Barrass, R. *Students Must Write: a guide to better writing in course work and examinations,* Routledge, London.

Barrass, R. *Scientists Must Write: a guide to better writing for scientists, engineers and students,* Chapman & Hall, London.

Gowers, E. *Plain Words* 3rd edn revised by Greenbaum, S. & Whitcut, J., H.M.S.O., London.

Glossary

To find the meaning of any term used in this book, look first at the Index. This glossary includes only terms not defined elsewhere in this book.

Abiogenesis (spontaneous generation): the theory, for which there is no evidence, that organisms can arise from non-living matter. *See also* biogenesis.

Alkali metals: group 1 metals (see page 146) which react with water forming an alkaline solution.

Amniocentesis: extraction of cells from amniotic fluid (see Figure 22.5) so that chromosomes can be examined to see if there are defects, or to find the sex of the developing baby.

Anatomy: the study of the structure of organisms (their organs and organ systems). *See also* histology.

Apparatus: equipment used in scientific investigations.

Applied science: the use of science in attempts to solve economic, medical and social problems. *See also* pure science.

Argument: logical reasoning which, starting from a correct premise, should lead to a correct conclusion.

Artificial respiration: misleading term. *Prefer* artificial resuscitation.

Asexual reproduction: reproduction without the formation and fusion of gametes.

Bar chart: diagram comparing two variables, one of which is not numerical (see Figure 29.1A). The blocks must be equal in width, and should not touch, but may be in any order.

Base units in the International System of units (see page 55): the metre, kilogramme, second, ampere, kelvin, candela, and mole. All other SI units are derived from these base units and so are called derived units.

Biodegradable materials: anything that can be broken down (degraded) by living organisms (especially by decay in soil).

Biogenesis: the *biogenetic law* is that *living organisms come only from pre-existing organisms* (by biogenesis).

Biological control: the reduction in population size of one species of organism by the activities of another species.

Birth rate: number of births in a year per thousand of population of a town, nation, etc.

Botany: the study of plant life.

Boyle's law: *the pressure of a fixed mass of gas, at a constant temperature, is inversely proportional to its volume.*

Cancer: uncontrolled cell division resulting in the formation of a tumour.
Calorie = kilocalorie = 4.2 MJ: the amount of heat needed to raise the temperature of 1000 cm^3 water by 1 °C. Note that this unit is no longer used by scientists.
Cellular respiration: misleading term. *Prefer* respiration.
Centripetal force: the force pulling towards the centre, which makes possible circular motion. Note that there is no centrifugal force.
Ceramics: materials made from fired clay.
Climatology: the branch of meteorology concerned with the mean physical state of the atmosphere.
Clone: genetically identical organisms resulting from asexual reproduction (without the formation and fusion of gametes), as in propagation from cuttings.
Cnidaria: phylum of animal kingdom (includes sea anemones, corals and jelly fish).
Colostrum: first milk secreted at time of birth.
Column graph: presents frequency distribution with discrete data, with blocks in order of increasing or decreasing magnitude which should not touch.
Community: (biotic): organisms of different species living in one place.
Competition: the use of one resource by organisms of more than one species.
Composite materials: mixtures, for example of sand, cement and water, which differ in properties depending upon the proportions of the ingredients.
Concept: an idea that has wide application (for example the concept of the atom, see page 39), which may or may not turn out to be correct.
Continuous variation: variation, usually quantitative differences between individuals (for example, in mass or height), such that the organisms do not fall into clearly defined groups. *See also* discontinuous variation.
Controversy: disagreement about the interpretation of evidence.
Cotyledon: seed leaf of a flowering plant (see Figure 5.4). See also dicotyledon and monocotyledon.
Death rate: number of deaths in one year per thousand of population of town, nation, etc.
Dialysis: separation of smaller molecules from larger molecules in a liquid by allowing them to pass through a differentially permeable membrane (for example, in the removal of excess salts and waste products of metabolism from blood in renal or haemodialysis in a hospital as a method of treating kidney disease).
Dicotyledon: broad–leaved flowering plants, with two seed leaves (see Figure 5.4) and leaves with branching veins (see Figure 15.2A), for example all vegetables except cereals. *See also* monocotyledon.
Discontinuous variation: variation due, usually, to qualitative differences between individuals (for example, differences in eye colour). *See also* continuous variation.

Domains: groups of atoms in a magnet, each group like a magnet (see page 301).

Doping: addition of traces of impurities to poor conductors of electricity to make them better conductors (see semiconductors, page 365). A semiconductor doped with arsenic has an excess of electrons (negative charges) and is called an n-type. But if doped with indium it has a deficiency of electrons and behaves as if it has an excess of positive charges — and so is called a p-type semiconductor.

Dynamo: a small generator that contains a permanent magnet (page 309).

Echinodermata: phylum of animal kingdom which includes starfish, brittle stars, sea urchins, sea cucumbers, and sea lilies.

Edaphic factors: soil conditions affecting, for example, plant growth.

Electronmicrograph: photograph of the image produced by an electron microscope. (see Figure 7.2B).

Electroplating: transferring metal ions from solution, by electrolysis, to the surface of another metal (for example, as a means of protecting iron or steel against rusting).

Endemic disease: disease that is always present in a population.

Epidemic disease: disease that occurs from time to time in a population, as a result of being brought in from outside.

Ethology: the study of the behaviour of animals, especially in their natural surroundings.

External respiration: misleading term, prefer gas exchange (see page 213).

Generalisation: general statement which provides a summary of many observations (for example, *all metals conduct electricity*).

Germ: not a scientific term. Used in everyday language for a microbe that causes disease (as in germ warfare) and for a gamete (as in germ cell).

Halogens: group 7 non-metals (see page 87).

Heterophytic and **heterozoic** modes of nutrition: unnecessary and misleading terms. *Prefer* heterotrophic (see page 264).

Histogram: presents frequency distribution with continuous data. The blocks, arranged in order of increasing or decreasing magnitude, should touch.

Histology: the study of the tissues of the body.

Incomplete dominance: when alleles of a pair both affect the phenotype, as in the inheritance of the AB blood group.

Incubation period: the time between infection and the appearance of the symptoms of a disease.

Inference: an interpretation of evidence, which is not necessarily correct.

Invertebrates: animals with nothing in common other than their lack of vertebrae. Misleading term, as this is not a natural grouping of organisms.

Irritability: sensitivity of organism to stimuli, which may be indicated by a response.

Joulemeter: instrument for measuring electrical energy supplied.

Larynx: voice box (part of trachea next to pharynx, (see Figure 18.5).

Light year: unit of measurement used for distances between stars (the distance light travels in one year = about 10^{12} km).

Line graph: a diagram in which changes in one measurement in relation to another are indicated by a line (as in Figure 5.2).

Locomotion: movement from place to place.

Lycopodophyta: club mosses (a phylum of the plant kingdom).

Medicine: the study of the symptoms and treatment of disease. *See also* pathology.

Metallic bonding: attraction of atoms in giant structures, both pure metals and alloys, in which electrons are free to move. Contrast with ionic and covalent bonding (see pages 143 to 144).

Mole: (a base unit in the international system): the sum of the atomic masses of a substance expressed in grammes.

Momentum: mass × velocity. The *law of conservation of momentum is that if no external force acts on a body moving in a particular direction, the total momentum of the body in that direction will not change.*

Monocotyledon: narrow–leaved plant, with one seed leaf and leaves with parallel veins (for example, the grasses — including all cereal crops).

Mutation: a change in the genetic material of an organism.

Nuclear fusion: the fusion of hydrogen atoms (atomic mass 1) to form helium atoms (atomic mass 4) in which energy is released, maintaining a temperature of about 15 million K at the sun's centre.

Oceanography: the study of seas and oceans.

Organelle: structurally distinct part of cell that has a particular function (for example, a chloroplast).

Oxidation: when a chemical substance loses electrons. See also page 143.

Pacemaker: concentration of nerve cells near heart, which controls heart beat (also electronic device used for this purpose as part of treatment for a medical defect).

Pandemic disease: disease which, starting in one place, spreads throughout the world.

Pathology: the study of disease.

Pest: an animal living where it is not wanted by people. The same species in another place may not be a pest.

Photomicrograph: photograph of image produced by a light microscope.

Physiographic factors: effects of lie of land on, for example, plant growth.

Pie chart: pictorial representation of numerical data, in which each segment of a circle is calculated as a fraction of 360° (as in Figure 18.3). The largest segment begins at noon, the next largest follows, and so on. If a second chart is drawn for comparison, however, the segments should be in the same order irrespective of their size.

Precipitation: rain, hail and snow fall. See also precipitate, page 116.

Predator: animal that kills and eats other animals (its prey).

Prey: animal killed and eaten by a predator.

Principle: a conclusion based on observation which provides a guide to what will happen in further observations.

Probability: the chance that something may happen (an estimate based on a sample of a population or on a calculation).

Proof: evidence that seems conclusive.

Protoctista: some biologists include the Protista and algae (see Table 7.1, page 73) and also the slime moulds in a kingdom called the Protoctista.

Protoplasm: misleading and obsolete term used when scientists thought there was a living material essentially different from non–living matter.

Psychology: the study of the behaviour of animals, especially in laboratory experiments. *See also* ethology.

Pure science: the pursuit of knowledge for its own sake. *See also* applied science.

Range: highest and lowest values.

Red blood cell: not in fact a cell, because it has no nucleus. Prefer term erythrocyte. *See also* page 222.

Relay: electromagnet that operates a mechanical switch; may act as an amplifier by using a small current to switch on and off a larger current.

Respiratory surface: a misleading term. Prefer gas exchange membrane.

Respiratory system: misleading term. Prefer gas exchange system (see page 213).

Saprophyte: a misnomer unless fungi considered plants (but see Table 7.1). Prefer saprobiont.

Saprozoite: a saprobiont which is an animal. See saprophyte. Prefer term saprobiont.

Secondary sources: review articles and books (as distinct from reports of original work, called primary sources, published in scientific journals).

Sphenophyta: horse tails (a phylum of the plant kingdom).

Taxonomy: the study of the principles and practice of biological classification.

Tissue respiration: misleading term. Prefer respiration (see page 213).

Transducer: structure that changes energy from one form to another, as does a sense cell (see page 227) or an electronic sensor (see page 366).

Transformer: electrical device in which an alternating current in one solenoid (the primary) induces an alternating current in another solenoid (the secondary). Depending on the number of turns in each coil the induced e.m.f. may be either larger or smaller than the voltage applied to the primary.

Transpiration: the loss of water by evaporation from the aerial parts of a plant.

Tropism: a movement of an organism towards or away from a stimulus; including growth responses of plants (for example, positive phototropism = towards light).

Ultrasound: high–frequency sound (above about 15 000 Hz), too high-pitched to be heard by people. See page 181.

Ultrastructure: structure too small to be seen even with a light microscope, and known only as a result of electronmicroscopy (see page 74).

Zoology: the study of animal life.

Index